VILLA D'ESTE

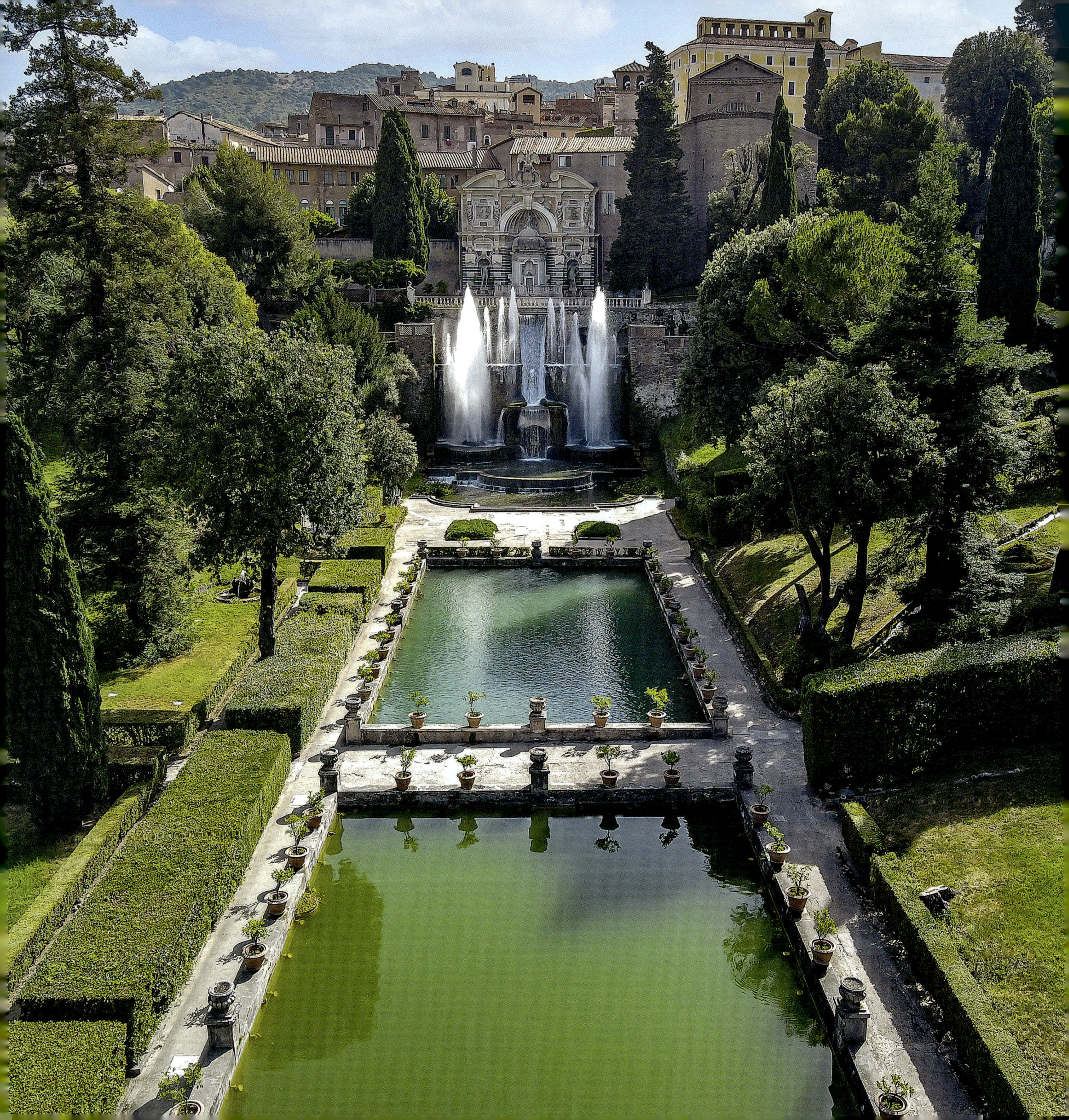

VILLA D'ESTE

Text by
Andrea Bruciati

SKIRA

Cover
View of the Fish Pools,
in the background are visible
the Neptune Fountain
and the Fountain of the Organ

Editor
Massimo Zanella

Design
Luigi Fiore

Editing
Giovanna Dalla Riva

Layout
Studio MZ, Milano

Translations
Lucian Comoy

First published in Italy in 2024
by Skira editore spa
Palazzo Casati Stampa
via Torino 61
20123 Milano
Italy
www.skira.net

Printed and bound in Italy. First edition

ISBN: 978-88-572-4760-1

Distributed in USA, Canada, Central & South America by ARTBOOK | D.A.P.,
75 Broad Street, Suite 630, New York, NY 10004, USA.
Distributed elsewhere in the world by Thames and Hudson Ltd., 181A High Holborn, London WC1V 7QX, United Kingdom.

Photo Credits
© Cristina Annibali/Krinaphoto: cover, 2, 36, 38–39, 44–45, 47, 78–79, 80, 86–87, 88–89, 90–91, 93, 94, 96–97, 99, 100–101, 102–103, 104–105, 108, 109, 110–111, 112, 113, 115, 120, 123, 124–125, 127, 128–129, 130–131, 132–133, 139, 140–141, 142–143, 144–145, 146–147, 148–149, 150, 152–153, 157.
© Getty Images: 10, 13, 14–15, 17, 20, 24–25, 28–29, 32–33.
© Istituto autonomo Villa Adriana e Villa d'Este-VILLÆ del MiC/Photo Quirino Berti: 6, 40–41, 48 sinistra, 48 destra, 49 sinistra, 49 destra, 52–53, 54–55, 56–57, 58, 60–61, 65, 66–67, 68–69, 70–71.
© Shutterstock: 51, 62–63, 76–77, 82–83, 85, 116–117, 118–119, 134–135, 136–137, 154–155, 158.

Acknowledgements
Laura Baruzzi, Angela Chiaraluce, Lucilla d'Alessandro, Rita Fabio, Giuseppina Morrone, Arianna Trifogli.

Special thanks to
Giuseppina Enrica Cinque

Contents

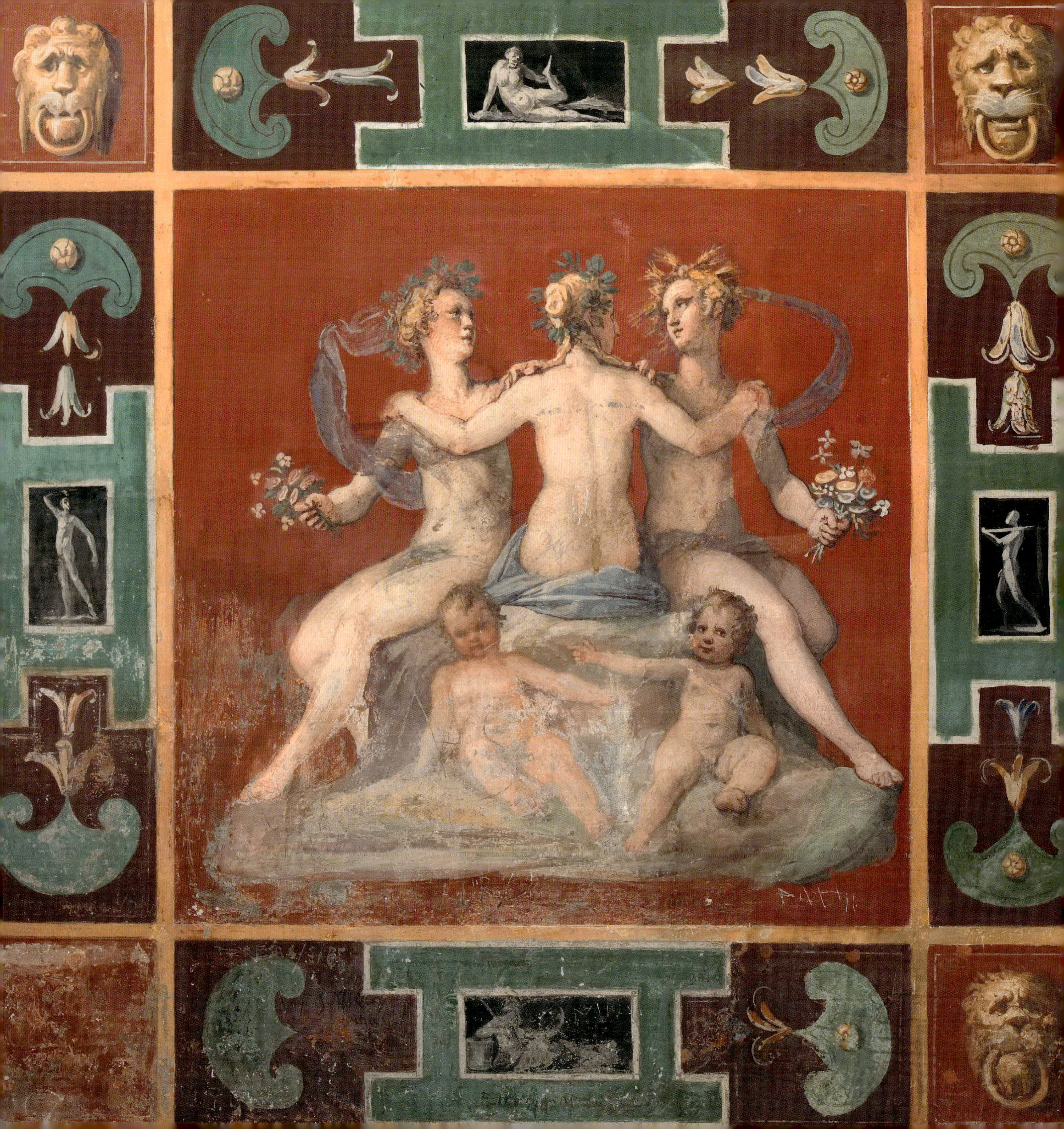

Introduction

Hadrian's Villa, a sumptuous imperial residence built from AD 118 onwards, and Villa d'Este, the cardinal's residence of Ippolito II d'Este, built from 1550 onwards, are both extraordinary and exceptional sites. Both villas constitute the material representation of a cultured and refined ideal of life; neither were ever completed nor did their initial owners ever thoroughly enjoy them; both were repeatedly plundered over the centuries, following the death of their creators, with the memory of their beauty finishing in museums all over the world.

The link between the two villas is close, as one draws its origin and inspiration from the other. After admiring the richness and uniqueness of the older villa's architecture and landscape and appreciating its close relationship between buildings, gardens, and water, the cultured and wealthy Este cardinal decided to reinterpret it in the light of Renaissance taste and aspirations.

Villa d'Este is, therefore, unique for this close relationship with the antique model: only in antiquity had such magnificence been achieved, a titanic feat that aroused the amazement of contemporaries, representing an absolute novelty in the panorama of sixteenth-century villas. Never before had such an astonishing number of fountains, grottoes, and nymphaea been concentrated in a single complex, covering little more than four hectares. The result was a marvellous place in which, as Umberto Foglietta wrote to Cardinal Flavio Orsino in 1569, "wherever you look, springs gush forth in such a variety of ways and with such splendour of design, that everywhere else on earth of this kind is far inferior".

Workshop of Federico Zuccari,
Friendship, 1566–1567,
Hall of Nobility, *piano nobile*

THE HISTORY

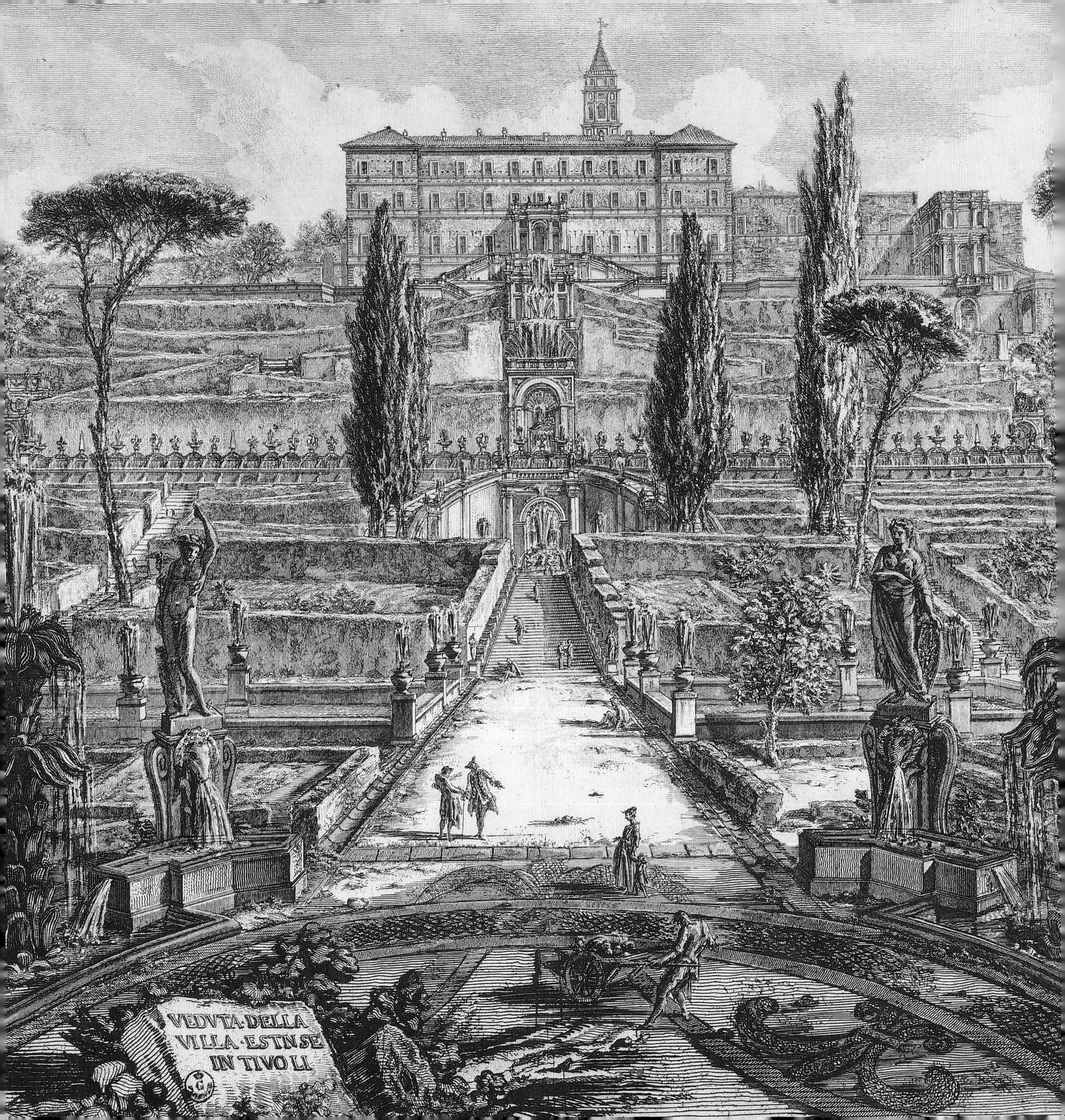
VEDUTA·DELLA
VILLA·ESTNSE
IN TIVOLI

Ippolito II d'Este and the villa

"Allo Illustrissimo & Reverendissimo Cardinal di Ferrara D. Hippolito da Este [...] per dare uno illustre testimonio delle magnifiche, et eccellenti fabriche, che ella ha fatto, & fa tuttavia in diverse parti del mondo con meraviglia de gli uomini: delle quali opere io ne haveva vedute alcune prima, che io le dedicasse il Vitruvio, alcune ho veduto dapoi, & sono quelle che con tanta splendidezza ella ha fatto in Roma & a Tivoli".
Daniele Barbaro, *I dieci libri dell'Architettura di M. Vitruvio*, in Venetia, appresso F. de' Franceschi e G. Chrieger, 1567

On 9 September 1550, the city of Tivoli celebrated the appointment of the new governor Ippolito II d'Este with a triumphal procession, as documented by the Tiburtine chronicler Giovanni Maria Zappi. Ippolito II was a member of a family member that had held a significant position within the nobility since the beginning of the first millennium. He boasted illustrious connections with the most distinguished European families, and these relationships endured over the centuries. Ippolito II was born during a period when his family was strengthening its alliances across Europe. He had close ties with France, as his sister-in-law, Renée di Valois-Orléans, was married to his elder brother, Ercole II, and she was also the sister-in-law of King Francis I. Additionally, the family had connections with the Church, as his mother, Lucrezia, was the daughter of the second Borgia pope, Alexander VI, and his illegitimate half-brother Alfonso had married Giulia della Rovere, the daughter of Francesco Maria, Duke of Urbino, and the niece of Pope Julius II. The family's affiliations extended to the Kingdom of Naples, as Ippolito II's grandmother, Eleanor of Aragon, was the daughter of the King of Naples. Moreover, it had ties with the ruling families of the Mantuan and Milanese regions. Ippolito's father had previously married Anna Sforza in his first marriage, one of his aunts, Isabella, was wed to Francesco Gonzaga, and another aunt, Beatrice, was married to Ludovico il Moro.

Giovanni Battista Piranesi, *View of the Villa Estense in Tivoli* (detail), ca. 1750

Born under the sign of Leo on 25 August 1509, Ippolito II embarked on the construction of the villa in Tivoli immediately after making an unsuccessful but close bid for the papacy. The prolonged conclave, which concluded on 8 February 1550, saw the election of Julius III (1550–1555). While his appointment as governor in Tivoli was a reward from Julius III for his support during the papal election, Ippolito II's wealth also derived from the numerous annuities granted to him by the King of France. He had spent three continuous years in France from 1536 and made several return visits, both for political reasons and to oversee the construction of his villa at Fontainebleau, known as the Hôtel du Grand Ferrare, a project he had commissioned from Sebastiano Serlio in 1542. It was probably during one of his stays in France that his daughter Renée (or Riniera, as found in older texts) was born. Sadly, she died in 1555, just two years after her marriage to Ludovico Pico di Mirandola. Their union was commemorated on two medals, now preserved in the British Museum in London, and celebrated in verse by Giulio de Berloni:

Renea d'Este anchor l'alma gentile
che di virtude al mondo ei tiene il vanto
Ne dir potria già mai con l'humil stile
della sua fama l'honorato canto
Ogn'altr'al mondo a lei si mostra vile
e tien di cortesia l'hornato manto
N pari a questa mai nel mondo sia
di virtù gentilezza, e cortesia.

The protection of Francis I enabled Ippolito to be the leading exponent of French cultural policy in Italy and, at the same time, assured him an annual income that made him the richest cardinal of his time, estimated at 120,000 scudi. However, this enormous sum

Jean Blaeu, *Hortus et Palatium Atestinorum Tyburi* (detail), in *Theatrum Civitatum et Admirandorum Italiae*, Amsterdam 1663

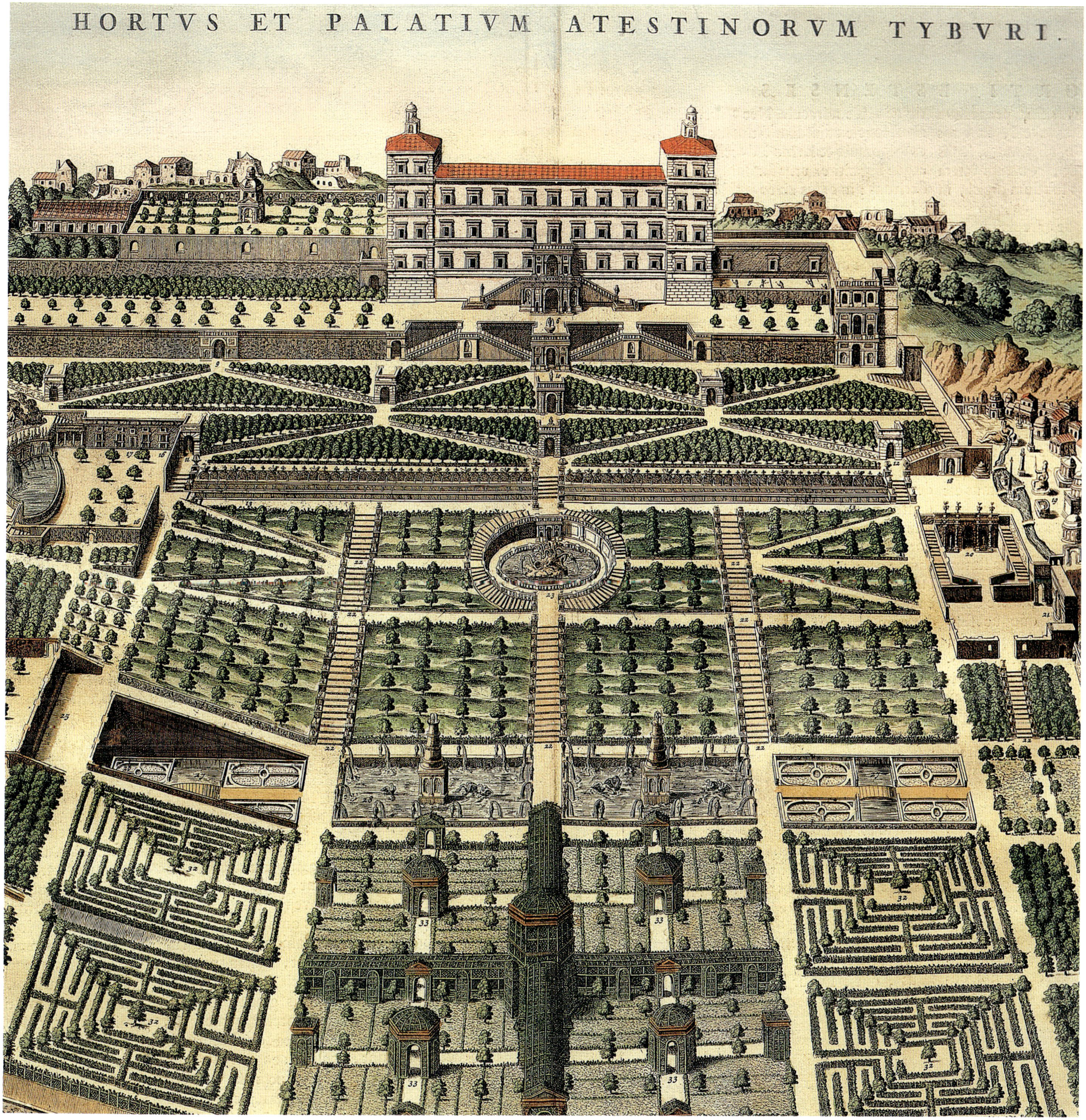
HORTVS ET PALATIVM ATESTINORVM TYBVRI.

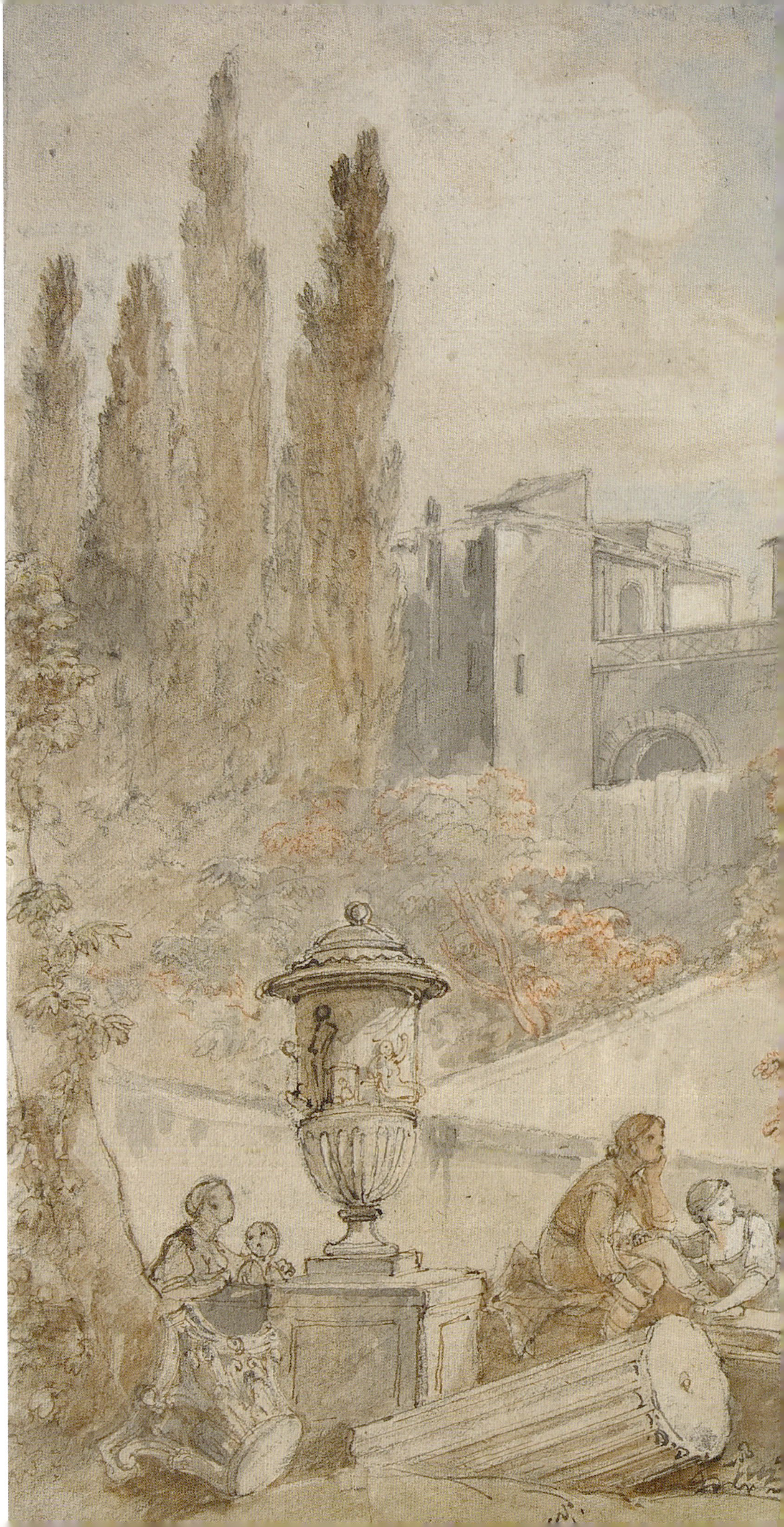

Charles Joseph Natoire, *The Gardens of Villa d'Este* (detail), 1760
New York, The Metropolitan Museum of Art

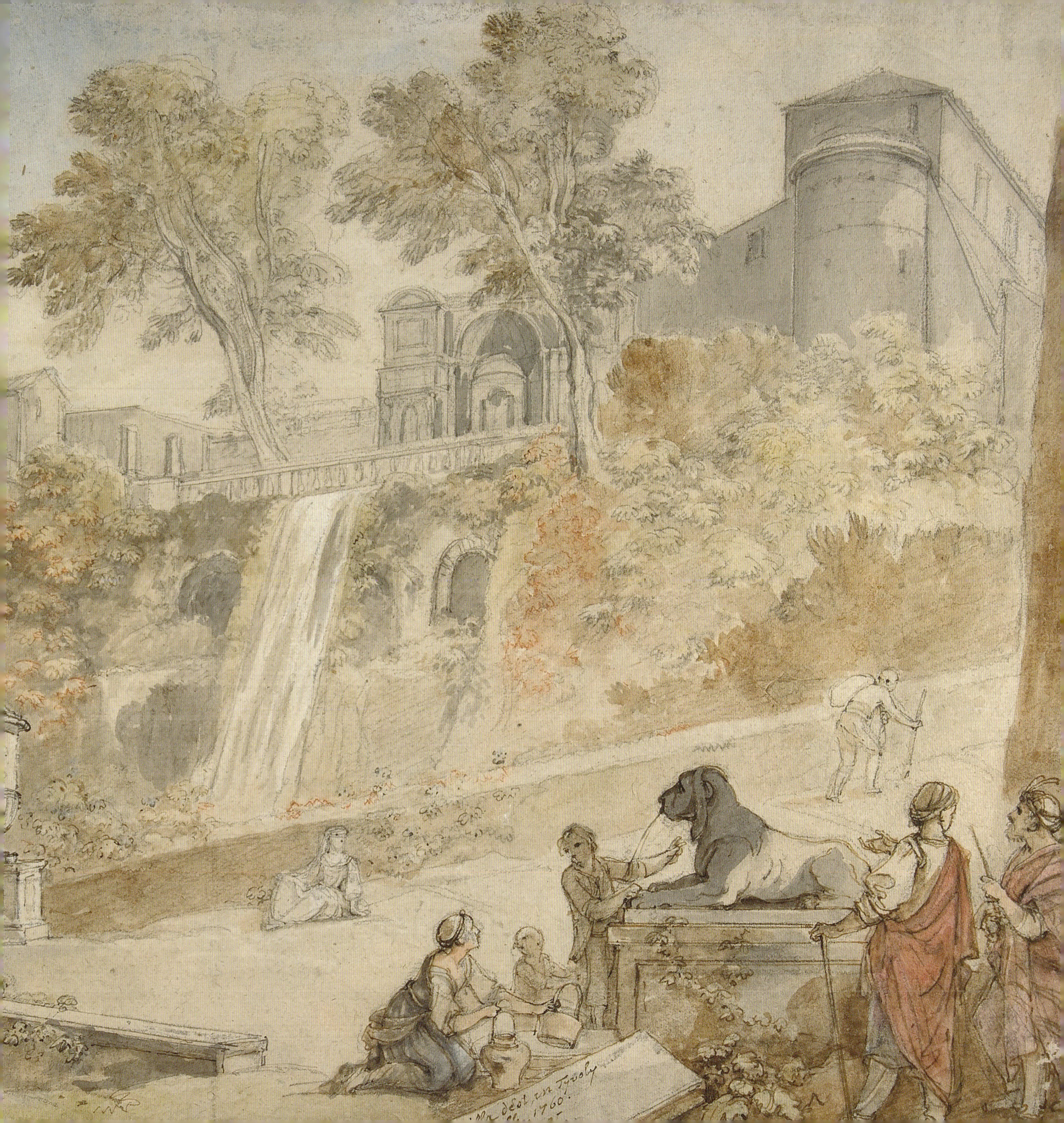
en Tyvoly
1760.

nevertheless barely covered his extravagant expenditures, leading to persistent pestering by creditors throughout his life. He was a munificent patron of numerous artists and men of letters who frequented his court, including Benvenuto Cellini, musicians Pierluigi da Palestrina and Nicola Vicentino, Latinists Uberto Foglietta and Marc-Antoine Muret, the architect Sebastiano Serlio, the antiquarian Pirro Ligorio and the poet Torquato Tasso. It is worth noting that the passion for architecture and for the harmonious integration of residences and gardens, the leitmotif of his dwellings, is a constant in Ippolito's comprehensive vision, and dates back to 1534 when he gained financial independence upon the demise of his father and began work on the functional reorganisation of the palaces he selected as his residences in the Ferrara district. Upon arriving in Rome for his papal candidacy, Ippolito promptly made a name for himself for extravagance: renowned as the richest man in the city, he organised feasts and banquets and, to his already established passions, began to add a particular interest in archaeology, initially expressed in the renovation of the Montecavallo garden (now the presidential residence of the Quirinale). This property, rented by the Carafa family in 1450, was later depicted in murals reflecting the modifications attributed to Ippolito, displayed in the Fountain Room of the Villa d'Este in Tivoli.

The first contacts with Pirro Ligorio – an architect of noble birth from Naples, who had moved to Rome and quickly gained recognition for his studies of antiquity. He had already begun to frequent and document the antiquities of the Tivoli district in 1538, back in the early years of his stay in Rome. It is therefore likely that before his arrival in Tivoli in his service as an 'antiquary', Pirro Ligorio had already embarked on some studies and survey of the territory, which continued until 1555. His investigations were wide-ranging, from excavations in search of statuary to studies of the architectural structures and installations of the villas. Notably, he was the first to create a planimetric reconstruction of Hadrian's Villa and provide descriptions of its individual buildings. Additionally, he delved into the geographical and morphological aspects of the sites. His work was accompanied by com-

Giannandrea Barotti, *Portrait of Ippolito II d'Este* (detail), in *Memorie istoriche di letterati ferraresi*, Ferrara 1777

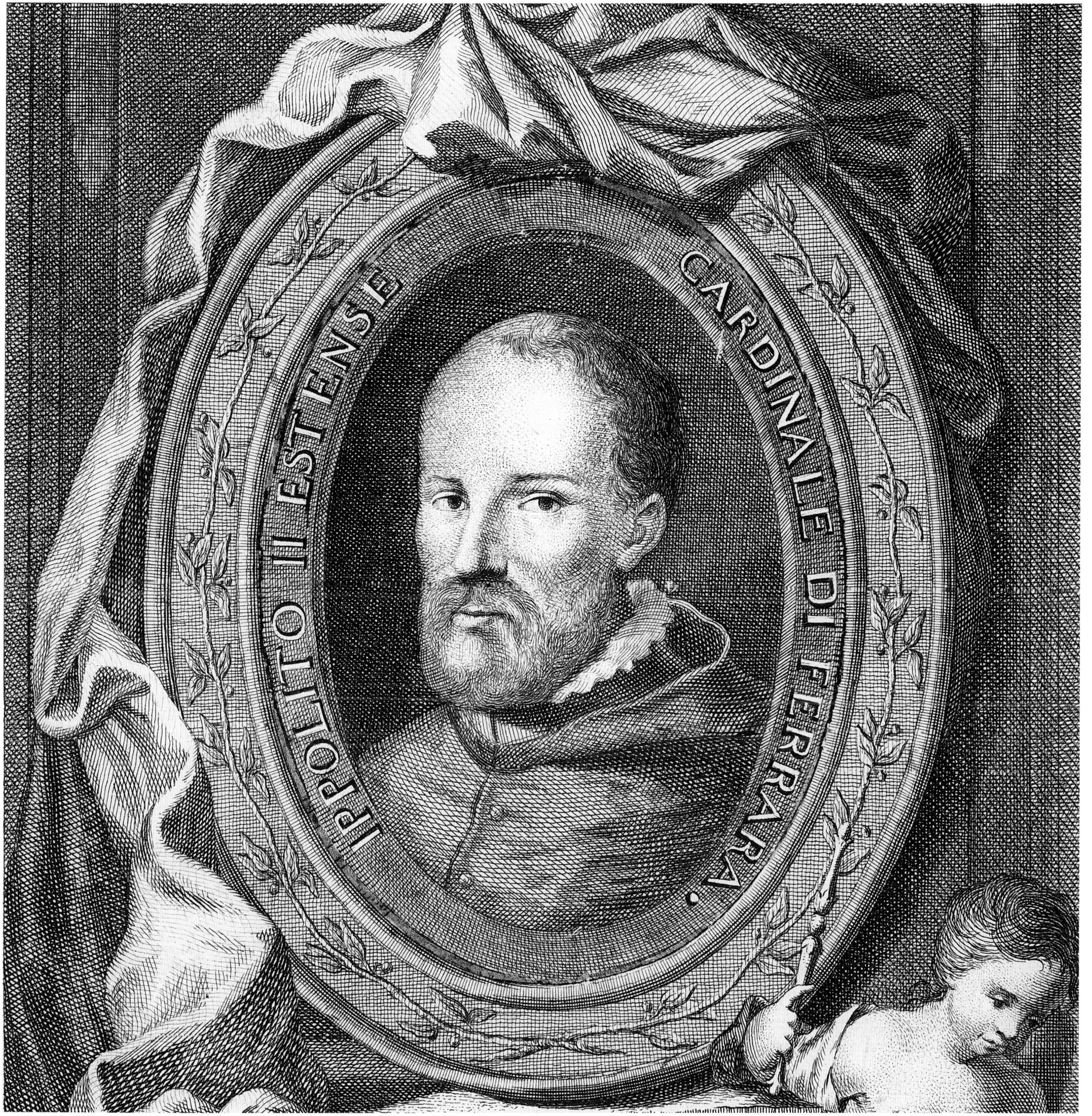
IPPOLITO II ESTENSE CARDINALE DI FERRARA.

ments, drawings, descriptions and reconstructive hypotheses: an encyclopaedic work later incorporated into the imposing manuscript work, the *Libri delle Antichità*. This work formed the foundation for the design of Villa d'Este.

During his initial sojourn, it was Ligorio who recognized the potential of the governmental seat's transformation into a "regal site", as noted by Zappi. It was also he who proposed the idea to Ippolito of establishing a garden of fountains in the estate's rear grounds, a concept that was both magnificent and ground-breaking. The north-east wing of the convent annexed to Santa Maria Maggiore was unsuitable for "housing such a prince" as Antonio del Re observed, despite previous work by the governors Bernardino Carvajal and Ercole Gonzaga. It also lacked the capacity to accommodate his extensive retinue. The monastery building did, however, enjoy a magnificent and commanding position and a wide view of the surroundings; according to the treatise that was being drawn up on gardens, these were fundamental requirements for the transformation of the palace into a villa.

Arriving in Tivoli in 1550, accompanied by a retinue of 250 followers, including more than eighty nobles and a multitude of artists, men of letters, philosophers and theologians who availed themselves of his protection, Ippolito II established himself in the palace previously occupied by his predecessors, later incorporated into the villa. Additionally, he rented a portion of the adjacent monastery of San Francesco. Ippolito wasted no time in initiating an extensive process of modernisation, encompassing both infrastructural and urban improvements, partly driven by his aspiration to create a residential environment worthy of his grand architectural vision. One of his initial undertakings included excavating an aqueduct to supply copious amounts of water from the Aniene River to the future villa. The ambitious architectural and urban planning endeavours led to the demolition of numerous houses and structures in the city, often sparking discontent among the local residents who were not entirely convinced of the benefits of these changes. However, the broader project aimed to rival the prominent Roman villas, particularly Hadrian's imperial villa on the

plain and Augustus's villa, located near Porta Romana, which would be adjacent to the future garden (Augustus's villa is the third site administered by the VILLÆ Institute, and corresponds to what is today called the Sanctuary of Hercules Victor).
The site chosen was a district called Valle Gaudente, which sloped steeply down from the foot of the monastery building towards Via del Colle at Porta Romana, the ancient entrance to the Via Tiburtina. Though part of this area lay within the medieval city walls, it still retained a semi-rural character, featuring scattered buildings amid vegetable gardens and vineyards, with a more developed area on the northeast slope toward the Campitelli district. Zappi described the place as being "of such a rustic and wild sort, with very large cavernous ditches, that one could freely say it was a home for wolves and other awful animals". Amidst the fervent construction activities in Tivoli, Ippolito also engaged in excavations at Hadrian's Villa, from which he recovered a substantial number of sculptures for adorning both Villa d'Este and the Quirinale. All activities were abruptly interrupted in 1555 when, amid struggles for the papal throne, Ippolito II was accused of simony and of a licentious lifestyle, and was obliged to flee for his life. His accuser, Cardinal Carafa, who would later be elected pope, always maintained a strongly hostile attitude towards the Este cardinal, revoking his governorship in Tivoli and seeking to diminish his influence through various means, including investigations by the Holy Inquisition, of which he as a staunch supporter. Ippolito II was only able to return to his villa four years later, following the death of the pope in July 1560. At that point, the area designated for the garden was still largely in private ownership, and so the campaign to acquire the land and demolish the pre-existing buildings, which lasted until 1569, was resumed. The first phase saw considerable efforts being dedicated to levelling the land, creating terraces with massive earthworks upstream and toward the Campitelli district, while the excess earth was moved to the right-hand side of the garden in the lower area towards the city walls and to fill depressions near the Porta Romana. At the same time, the artificial terraces were supported by means of *all'antica*

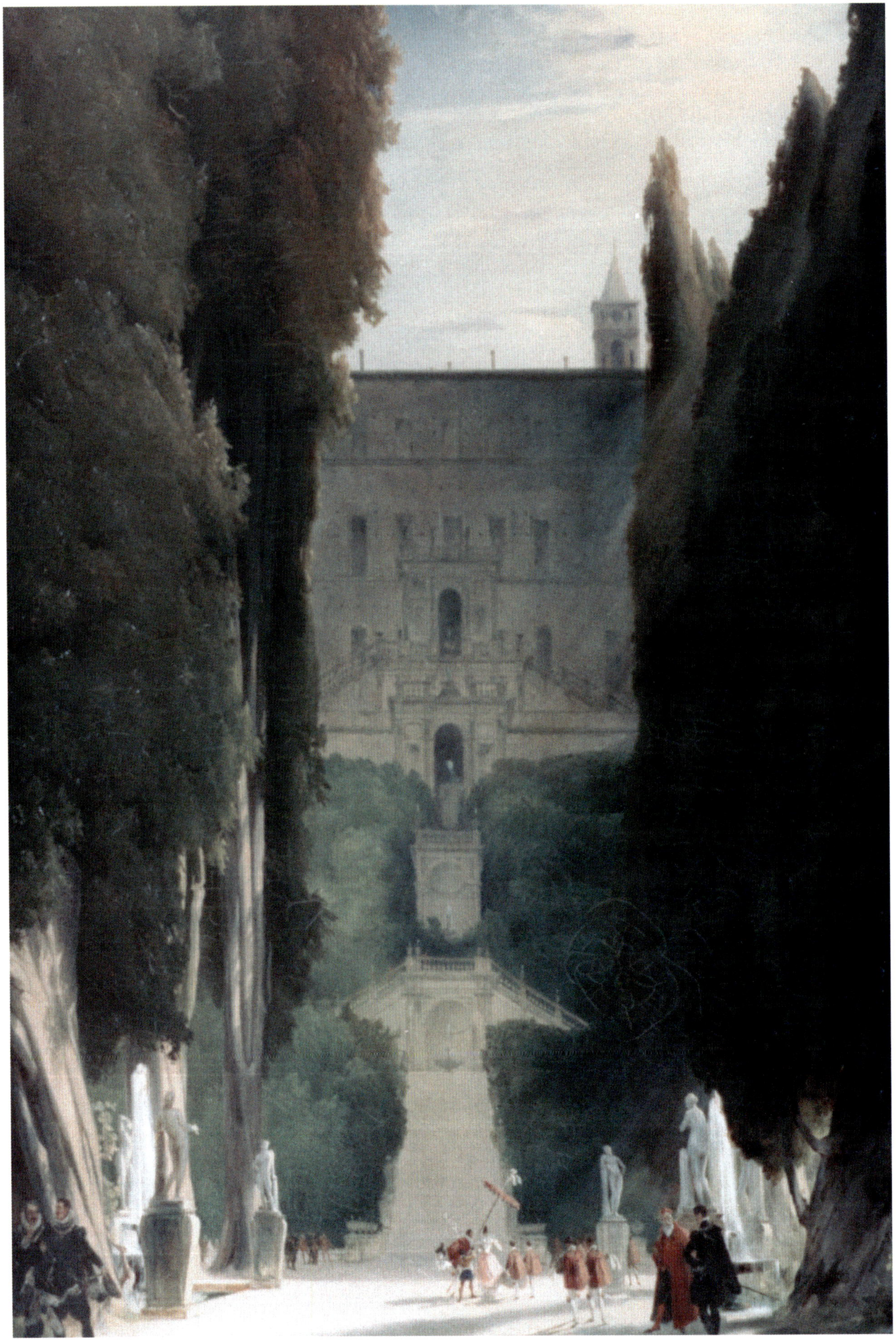

substructures ('in the antique manner') with strong pillars connected by arches or barrel vaults, creating vaulted chambers that were then used to create numerous grottoes, niches and 'chambered' nymphaea. At the same time, the garden's design was developed, aligned along a grid of orthogonal paths that subdivided the area into regular compartments, each a module measuring about thirty metres. These spaces were symmetrically organised concerning the longitudinal median axis, serving as the visual reference line, while five main axes in the transverse direction governed the architectural composition, offering unique lines of perspective toward the fountain backdrops.

Over the next twelve years and up to his death on 1 December 1572, having regained his position as governor of Tivoli and sparing no resources in his unsuccessful attempts to attain the papal tiara, he spent much time in the villa, seeing to its completion in accordance with an erudite iconological programme. This was drawn up by Pirro Ligorio and the humanists of Ippolito's circle. The goal was to ennoble the residence, enhance the cardinal's image, magnify his virtues and lineage, and imbue the architectural project with courtly meanings. The symbolic, allegorical and celebratory themes were integrated into all the elements of the architectural composition, from the many fountains to the paths and water features, from the arrangement of the vegetation to the pictorial decoration of the interiors, to the decoration of antique statues. The programme, based on a deep knowledge of the legends and themes of classical culture, was well understood in its semantic stratification by an intellectual such as Ippolito, who was accustomed to the symbolic language of mythology. After 1569 there was a slowdown in work, probably due to the unsettled financial situation caused by the huge expenses, but although suffering from gout, the cardinal stipulated in his will that the property should be inherited exclusively by Este cardinals, and in their absence, should be bequeathed to the Dean of the College of Cardinals. This decision was probably adopted to promote the candidature of a member of his family to the papal throne. However, this decision would later contribute to the villa's decline, as it was pillaged during periods without Este cardinals.

Carl Blechen, *The Gardens of Villa d'Este*, 1830
Berlin, Staatliche Museen, Nationalgalerie

While it is documented that the villa was designed by Pirro Ligorio, very little of what remains today corresponds to the work of this renowned sixteenth-century architect and antiquarian, as depicted in Étienne Dupérac's widely known 1573 engraving. The most significant alterations are found in the garden, which was never completed by Ligorio. After his appointment as papal architect in 1558, he returned to work for Ippolito II from 1560 until his subsequent appointment as the architect for all buildings at St. Peter's, succeeding Michelangelo. Later, between 1568 and 1569, he entered the service of Alfonso II d'Este and moved to Ferrara. Upon Ippolito's death, much of the antiquarian collection collected by the cardinal was sold at auction by his nephew, the new duke, Alfonso II. On inheriting the villa, his second son, Cardinal Luigi d'Este (1538–1586), entrusted the architect Flaminio Ponzio with completing the work. The first significant period of looting occurred after Cardinal Luigi's death, when, due to the absence of Este cardinals, the property passed to the Dean of the College of Cardinals. Local historians report that even garden vases were taken by Roman cardinals during these years. In 1604, Cardinal Alessandro d'Este (1568–1624), the illegitimate but recognised son of Alfonso d'Este (in turn the natural son of Alfonso I, that is, Ippolito II's father), who had just been appointed governor of Tivoli, succeeded in regaining possession of the villa and successfully concluded the lawsuit for unconditional possession by the Este family. During the twenty years that he spent in Tivoli, a city that, like the two previous Este cardinals, he chose as his burial place (the remains of the three cardinals rest in the church of Santa Maria Maggiore, adjacent to the entrance to the villa), he dedicated himself intensively to repairing the damage suffered during the previous twenty years or so, as well as transforming many parts of the garden with modifications to areas and the insertion of new fountains. His efforts were followed by interventions commissioned by his successor, Cardinal Rinaldo I d'Este (1618–1672), nephew of Luigi's brother, Cesare, the second son of Alfonso III and Isabella of Savoy and, therefore, brother of Duke Francesco I.

Under Rinaldo I, Bernini made several alterations in the garden, and he is remembered for having created a fountain, possibly the Bicchierone, using mud and wooden planks. In the subsequent years, thanks to the interest of Duke Francesco I, restoration work in the villa continued. These activities, however, came to an abrupt halt with the death of the duke, whose heirs, related to European ruling houses, neglected the property. Divided into apartments, the villa was often rented out, primarily to foreigners. One such tenant was the Scottish painter Allan Ramsay, who in 1755 had Robert Adam as a guest; Adam would later become the greatest architect of the period in Britain.

In 1796, upon the demise of the last duke of Este, Ercole III, the villa came under the ownership of his daughter Beatrice, who was married to Ferdinand Karl, Archduke of Austria, whose offspring adopted the name Austria-Este. Subsequently, the villa remained unoccupied for a period but was later taken over by French troops. It regained attention when Cardinal Gustav Adolf von Hohenlohe-Schillingsfürst, an aristocrat with a prestigious lineage distantly related to the Este family, expressed interest in the property. He secured its use in 1850 and dedicated himself keenly to its restoration. During his tenure, the villa played host to numerous artists and musicians, including Franz Liszt, between 1865 and 1885.

With the death of the Cardinal in 1890, the house of Austria-Este regained possession of the villa, despite their repeated attempts to dispose of it over the years. Such efforts consistently met resistance from the Church. It was not until the end of the First World War, with the confiscation of property from the defeated enemy, that the Italian State successfully acquired the villa. During the initial period of state ownership, extensive and, at times, intrusive restoration work was undertaken, particularly in the garden. This included the complete refurbishment of several fountains and the introduction of numerous pressurised water supply lines. Notably, the Neptune Fountain, whose current form is celebrated with a portion of a painting in the Hall of Fountains or Views, underwent significant modifications from 1920 onwards.

Pages 24–25
Giovanni Francesco Venturini, *View of the Palace from the Garden Plan with its Fountains*, in *Le Fontane del Giardino Estense in Tivoli*, Rome 1691

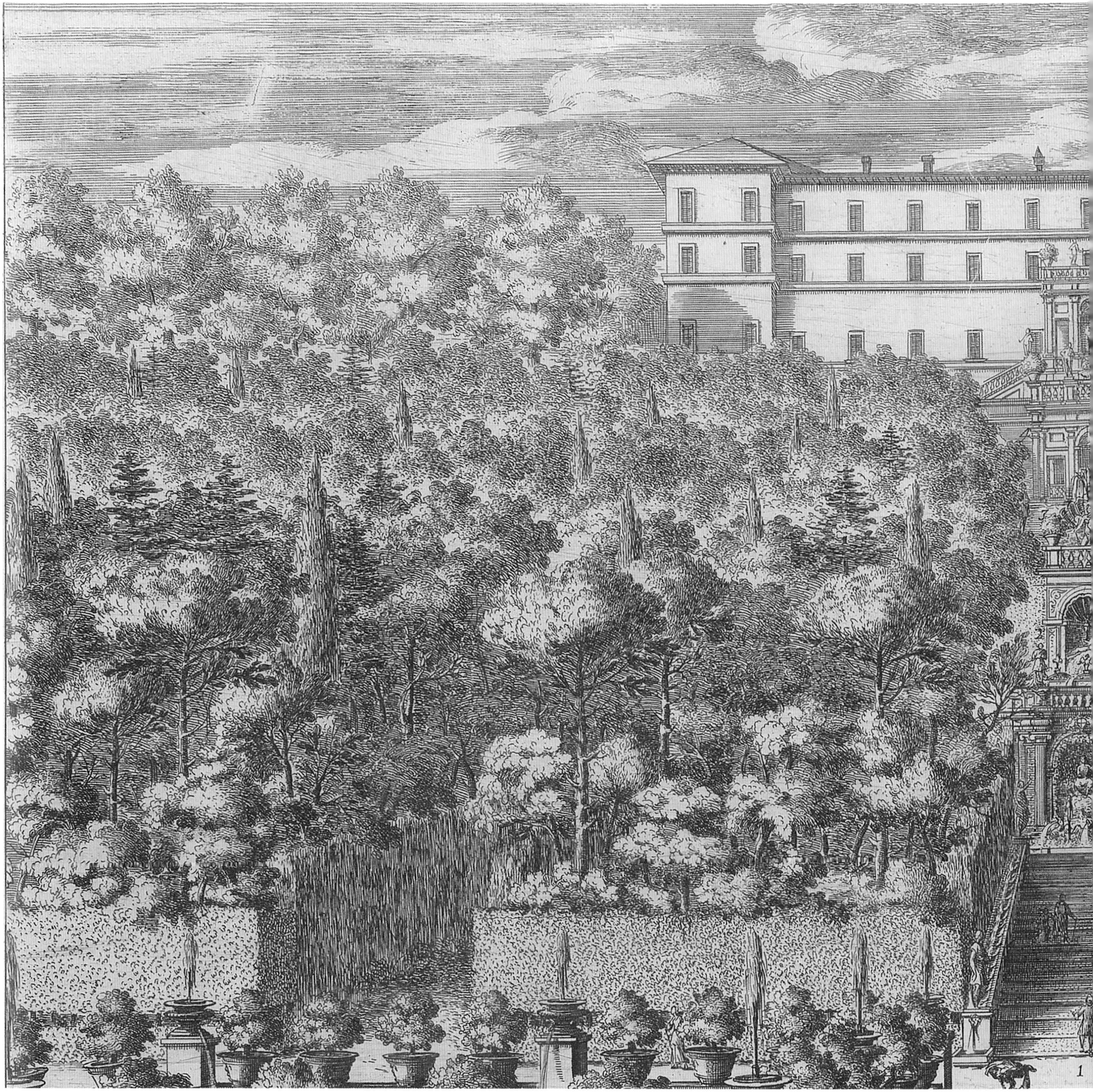

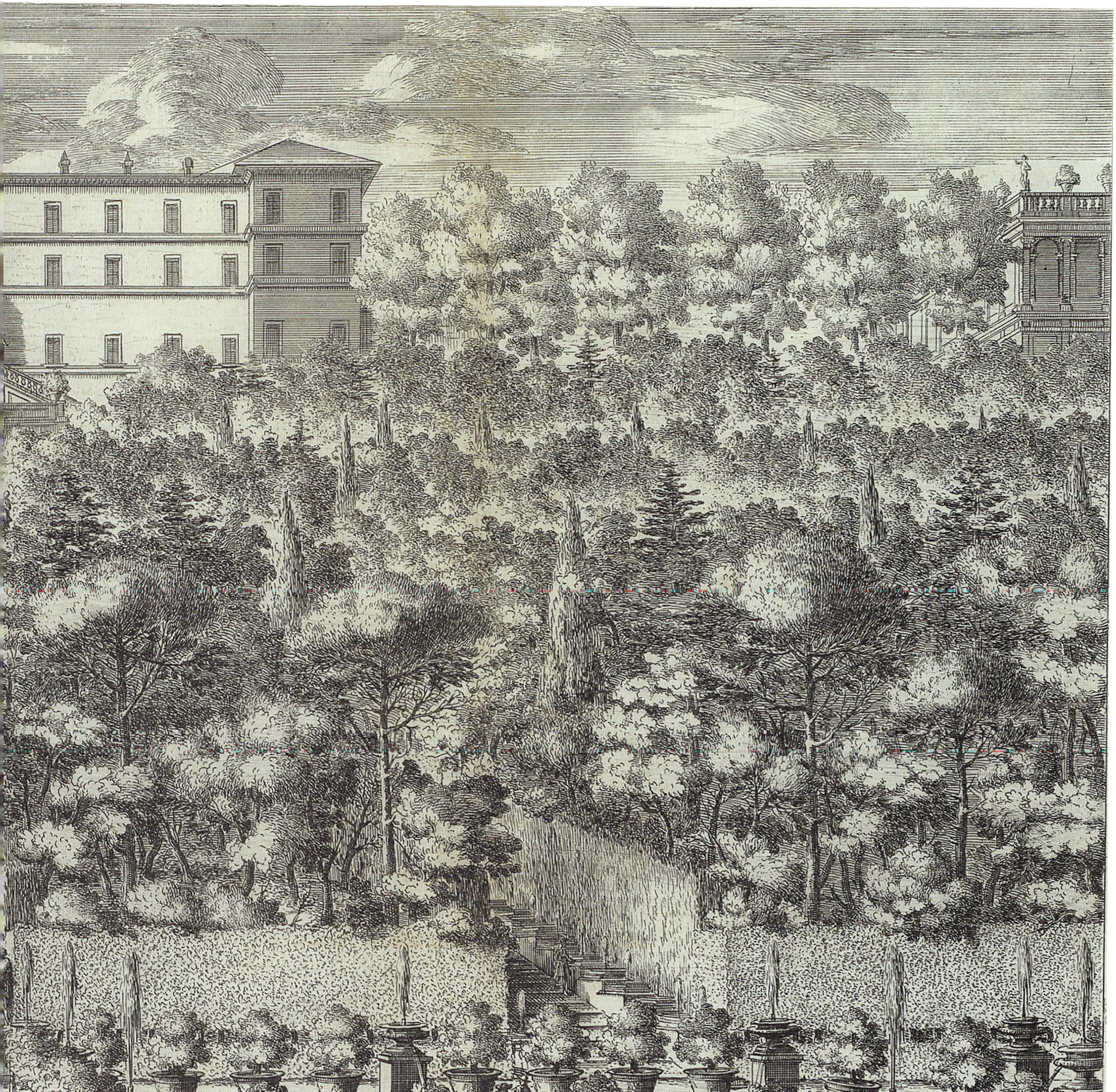

The antiquarian villa

Was Villa d'Este designed to evoke the spirit of Hadrian's Villa, intended as the abode of a cardinal who, having missed the opportunity to ascend to the papacy, envisioned himself as an illustrious emperor? Ligorio's Turin codex bears witness to his remarkable archaeological undertaking, involving an exhaustive exploration of the Tivoli region through meticulous source analysis, extensive excavations and detailed surveys. Ligorio diligently identified all the archaeological remnants within the vicinity, giving particular significance to the area associated with Augustus' villa (now the Sanctuary of Hercules Victor), renowned for its impressive ruins and formidable substructures nestled in the Aniene Valley, regarded at the times as an imperial villa, and to Horace's villa, located near Tivoli's cathedral. Consequently, the selected site was essentially a part of the expansive terrain demarcated by these two villas. Ippolito's villa and Ippolito himself could thus be associated with Augustus and his villa even before referencing other aspects of antiquity, which later found expression through the excavations of Hadrian's Villa.

The comprehensive exploration of the Tivoli district sought to establish a profound connection between the new villa and an archaeological context that would bestow upon it a sense of nobility: a territory embodying the self-representative image of the most eminent and opulent cardinal, a reincarnation of the imperial elite who once resided here. In his portrayal of Augustus' villa, Ligorio affirmed that the fundamental objectives of his archaeological research were to furnish models "for the benefit of those building in the villa [Este]". Additionally, the cryptoporticus on the lower level of the palace was constructed atop the walls of a magnificent late-Republican Roman villa (of which a decorated mosaic from the first century AD remains visible), potentially the same villa of Augustus erroneously identified in the Sanctuary of Hercules Victor, and extending into the garden. References to both Augustus and Hadrian were blended in a manner unique for its time, echoing another wonder of the world: the Hanging Gardens of Semiramis in Babylon. Ligorio lauded the *regina villarum*, stating, "Hadrian's villa was grander than any of the world's seven wonders, any building, and anything constructed thereafter. It was a marvel to be revered by everything fashioned for grandeur, magnificence, ostentation, or the ennoblement of cities, or for any commendable and virtuous endeavour. Its adornments were of an unparalleled grandeur, and never before had so many commendable elements been amassed in any century. May they rest, the labyrinths of Knossos, Egypt, and Tuscany, the Colossi of Rhodes and Egypt, and the other great buildings in Dodona and Cyzicus [...], for this villa, even though it may be fractured and in in ruins, still radiates splendour through the remnants of the ancient prosperity that building offered in honour of the Aelia family. To this, our most Illustrious and Reverend Messer Ippolito of our era, the second Este cardinal, has been able to contribute significantly through his extensive efforts in excavating these sites, unearthing a multitude of remarkable treasures". The same sentiment applies to the Villa d'Este: "There is no visitor who comes to see it, who does not marvel and become lost in its grandeur, and there is no one who is not confounded upon seeing it: such majesty is exquisitely embodied within it".

Hydraulic engineering

The abundant water resources, made possible by the remarkable feat of capturing the Aniene River, opened up a world of possibilities at Villa d'Este. It allowed for the first comprehensive exploration of the dynamic potential of water, enabling experimentation with various aesthetic and visual effects. To accomplish this, the construction of a complex 'water machine' was essential. It had to supply approximately 51 fountains and nymphaea, featuring 398 spouts, 364 jets, 64 cascades and waterfalls, 220 basins, tubs, ponds, and 875 linear meters of water chains and channels. This intricate network of tunnels and underground pipes represented an avant-garde engineering marvel for its time.

The idea for this project originated from Pirro Ligorio, inspired by the extraordinary subsoil characteristics of Tiburtine, renowned since Roman times for its abundant water sources.

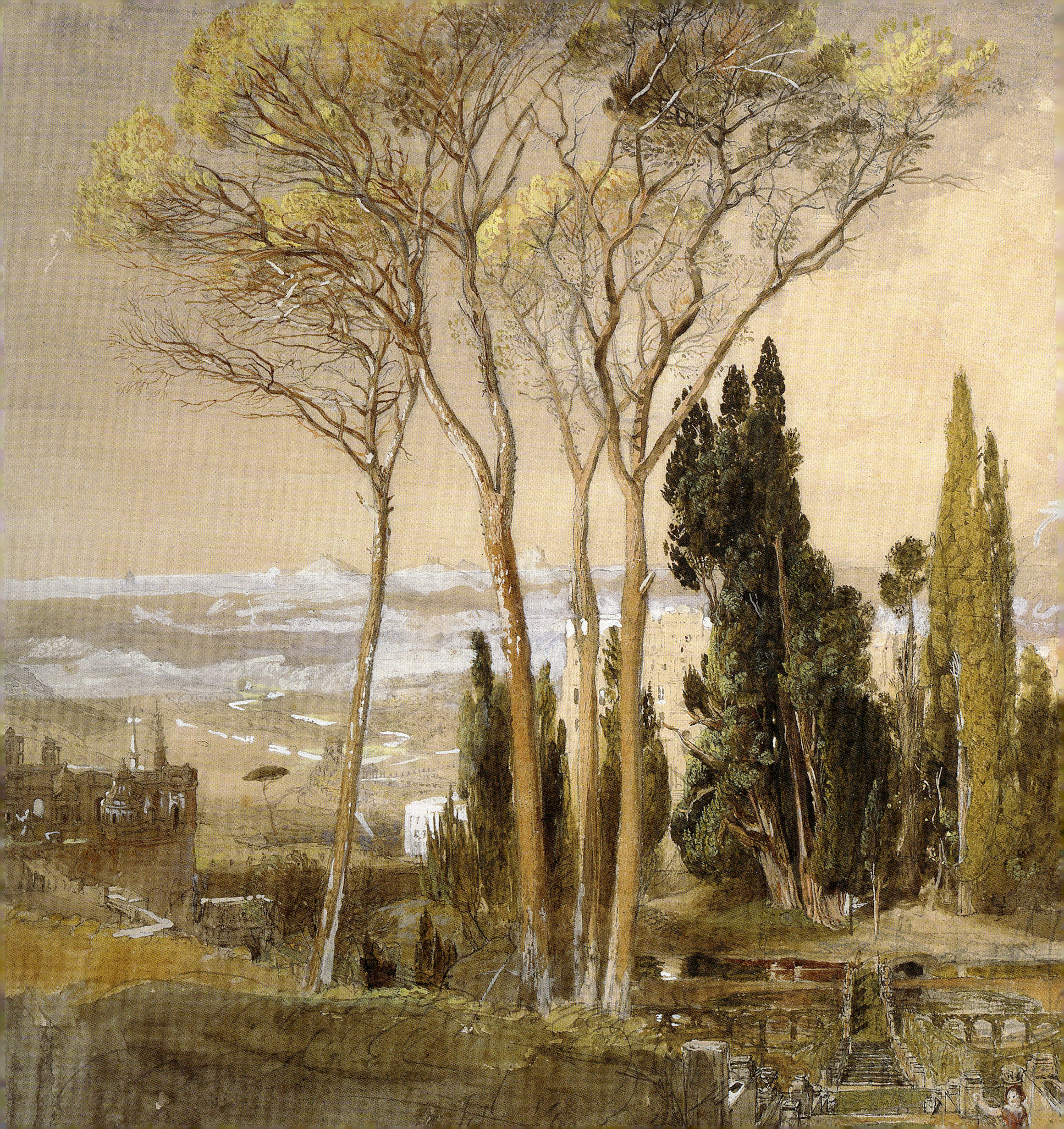

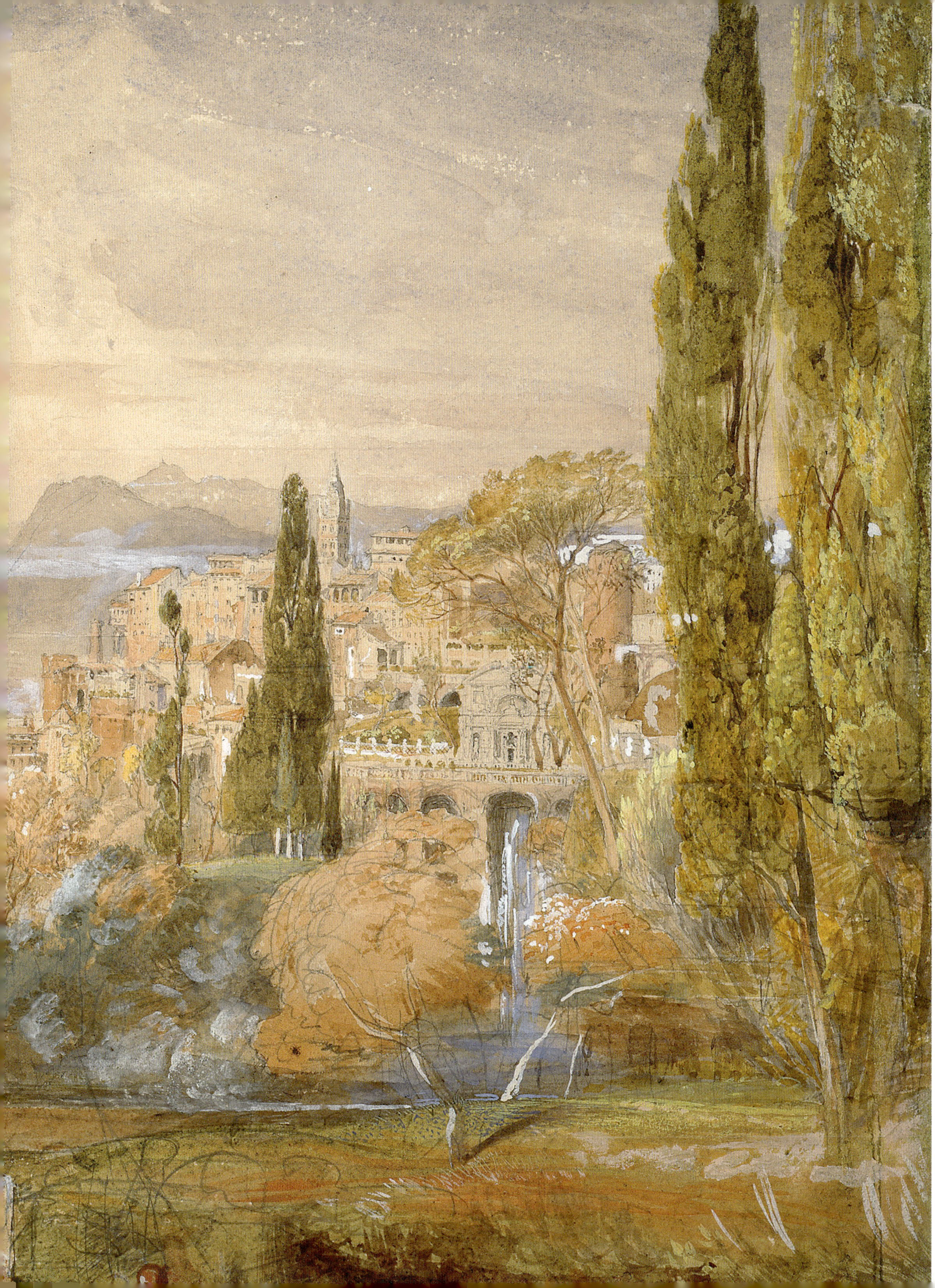

Samuel Palmer, *The Gardens of Villa d'Este* (detail), 1837
London, Victoria and Albert Museum

Apart from the primary Aniene river branch, which formed a stunning waterfall at the acropolis's base, secondary underground branches, collectors, tunnels, and passages contributed to picturesque small waterfalls along the limestone terrace where the city stood. The Santuario di Ercole Vincitore, believed to be Augustus's villa during Ligorio's era, added a final scenic touch before reconnecting with the Aniene.

The ancient Roman contributions, on which the water system was based, played a pivotal role. This knowledge was derived from the treatises of Vitruvius and Frontinus and likely from excavations at Hadrian's Villa, which Raffaello da Sangallo regarded as a rich source for defining various structures, from baths to nymphaea to ornamental basins, which used water during the imperial age.

The layout was designed to optimise water usage, allowing one fountain's discharge to supply those at lower levels, resulting in significant water savings and simplification of the water network. The initial water supply, established around 1560–1561, involved the Rivellese aqueduct, which fed the public fountain in the square in front of the current entrance. It divided into three branches: one entered the palace courtyard, while the second fed fountains in the palace and the upper garden. The third branch supplied the Fountain of the Unicorn in the Secret Garden. The Rivellese aqueduct's modest flow rate of five litres per second necessitated cisterns and reservoirs and limited water consumption to specific periods. In 1564–1565, a much larger conduit was constructed to draw water from the Aniene, providing an unlimited water supply, currently at five hundred liters per second. The water enters the villa upstream of the Fountain of Ovato, with distribution chambers distributing it to various outlets and outflow channels for agricultural and industrial use outside the garden.

The current system therefore preserves the general distribution and drop pattern of the sixteenth-century system designed by Tommaso Ghinucci, albeit with the progressive replacement of both the materials used and the supply source itself. By the end of the nineteenth century, the Rivellino aqueduct was effectively exhausted, and in the 1930s, it became necessary to resort again to the Aniene, which was raised from its point of entry to the cistern under the cloister, from which the distribution pipes branch off in line with the ancient layout.

The engineering function is at the service of the aesthetic enjoyment of water, which also serves as the primary architectural element and is moulded into innumerable forms, from numerous rivulets to simulated rain, sun rays, boiling water, or even a menacing deluge. The gardens featured pioneering elements like 'water theatres', 'water staircases', and the first artificial waterfall. The design envisaged the use of water combined with air to create the sounds of birds and animals, or of bombards, blunderbusses, trumpets, buccinas, or organs. These auditory experiences were made possible through engineering tricks influenced by the Hellenistic tradition, particularly hydraulic organs and automata.

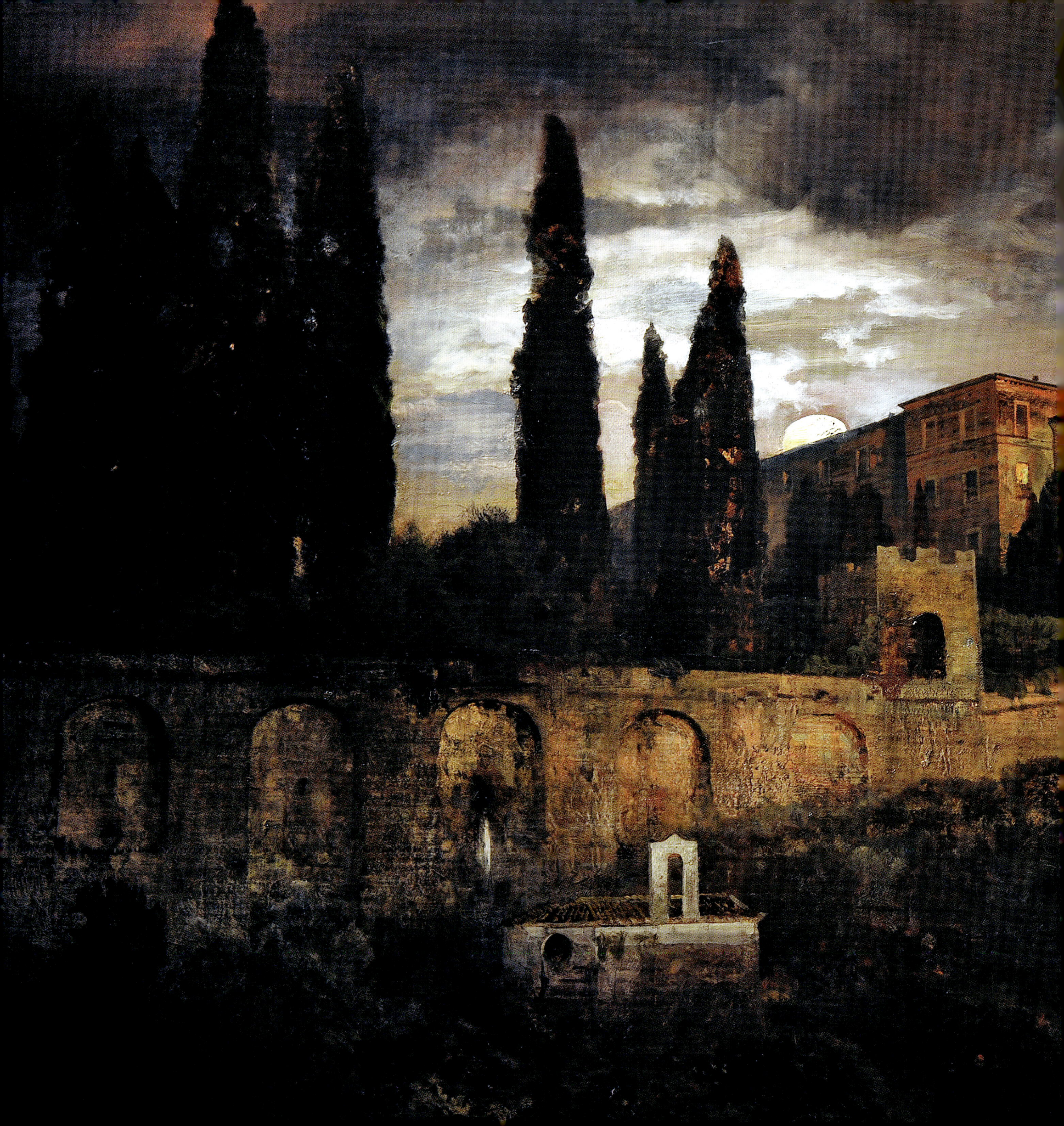

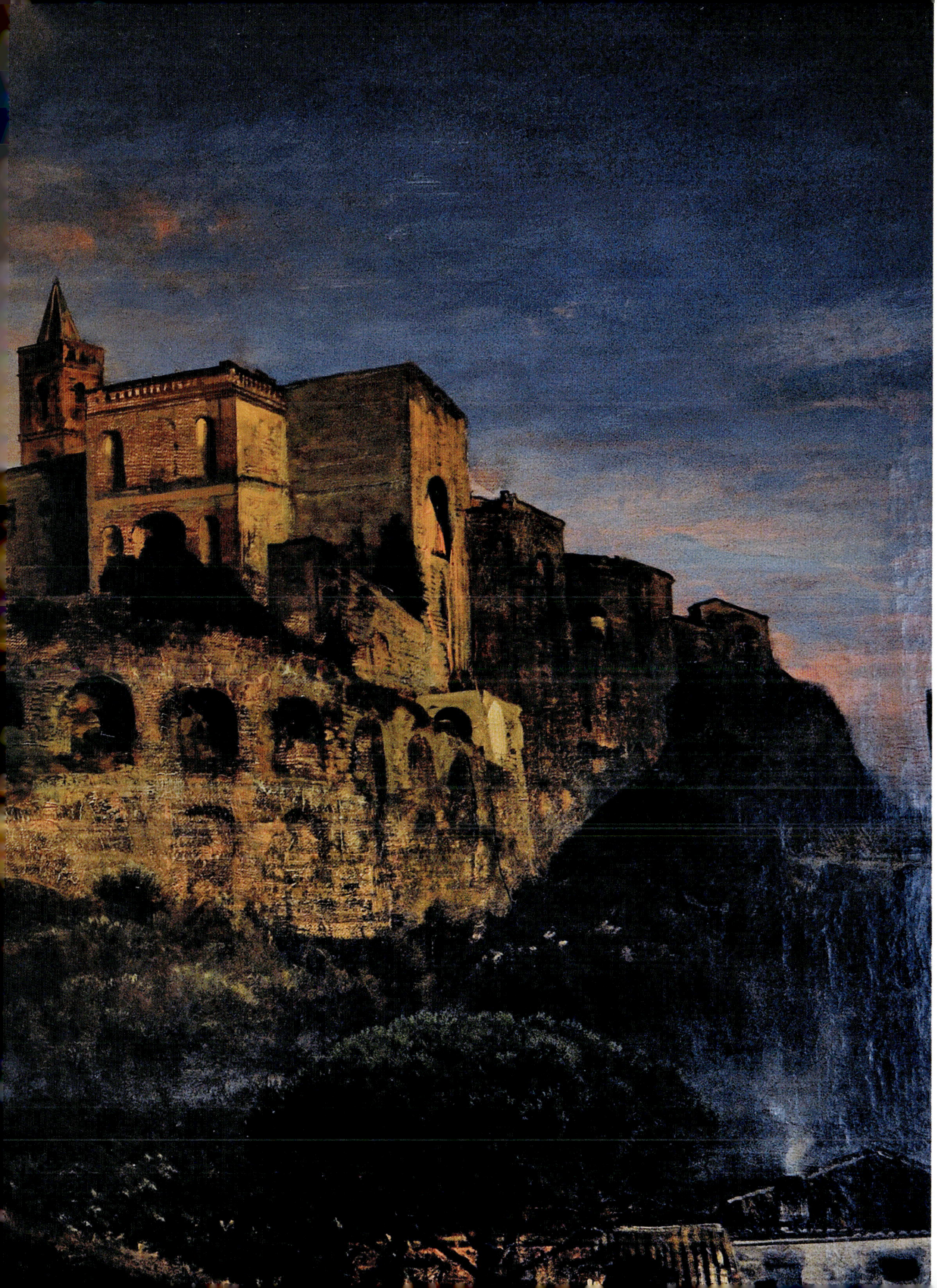

Samuel Palmer, *View of Villa d'Este* (detail), 1892
Budapest, Szépművészeti Múzeum

THE VILLA

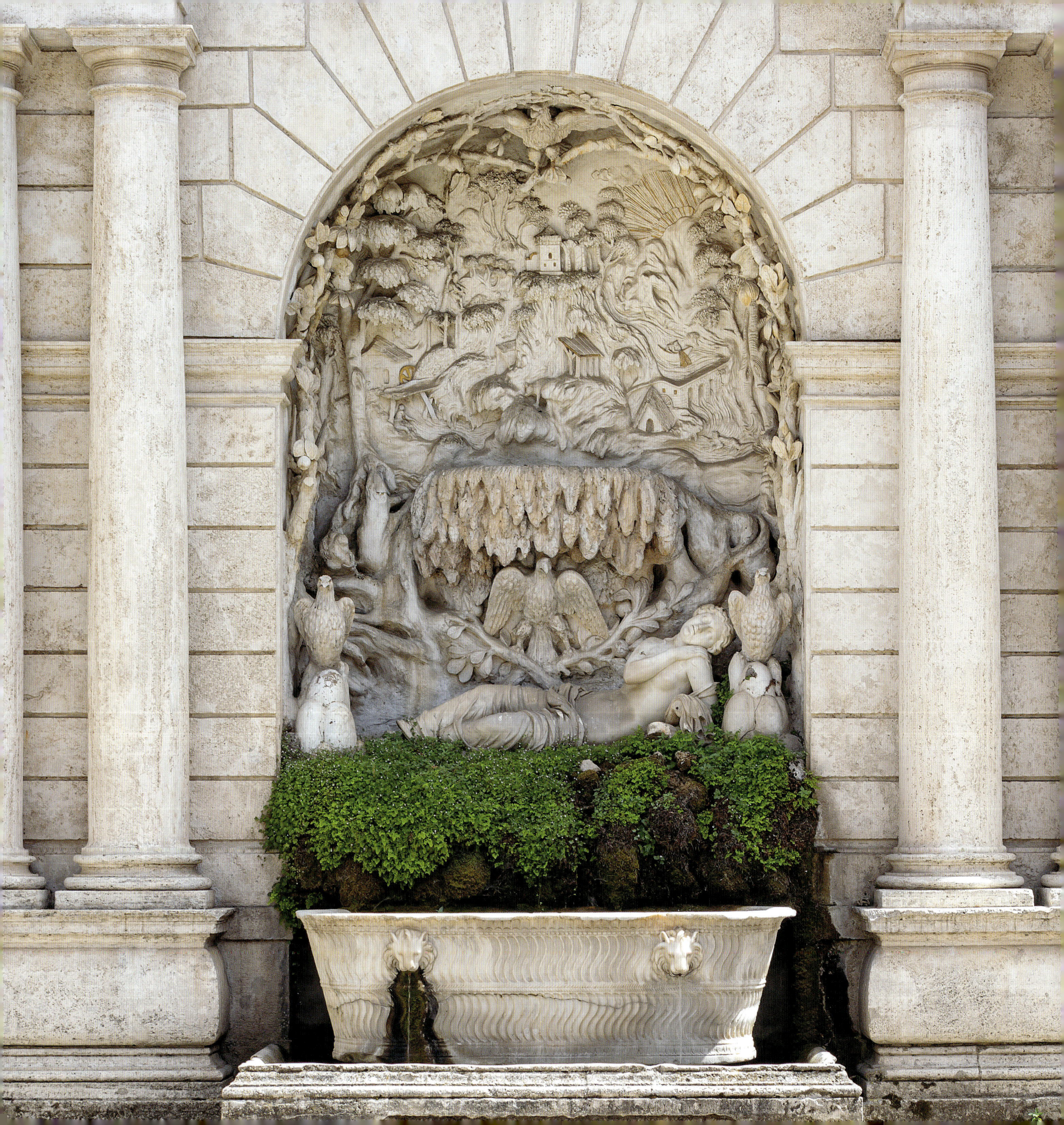

Palace entrance

The entrance gateway displays one of the few remnants predating Ippolito's interventions: the coat of arms of Cardinal Bernardino Carvajal, who served as the governor of Tivoli until 1521, adorns the arch's keystone. Passing through the portal, one enters the Androne (entrance), decorated with stories from the Old Testament. This room shows signs of the damage caused by Allied bombing during the World War II; the destruction and subsequent reconstruction of the north-east wing of the palace resulted from this.

Courtyard

Formerly the cloister of the pre-existing monastery integrated by Ippolito II, the courtyard displays few remnants of its former layout, except for a mullioned window in the northeast wall. The portico, featuring Doric pillars, round arches, and cross vaults, is the work of Raffaello da Sangallo, who is mentioned in 1565 for his work as a plasterer and in 1586 as an expert in hydraulics. He also designed the Fountain of Venus. Positioned against the wall of the Santa Maria Maggiore church, it is the only element that has preserved its original sculptures. This includes a slumbering Venus from the Roman era occupying the central niche, as well as a fourth-century bust portraying Emperor Constantine above the same niche. There is a vase from the second century beneath the niche to collect flowing water. The entire nymphaeum's design revolves around the theme of a triumphal arch, defined here by a single archway supported by pairs of Doric columns. A stucco bas-relief in the background depicts the course of Tivoli's Rivellese aqueduct, commissioned by Ippolito II to supply part of the villa. The bas-relief is encircled by a festoon of quince branches, alluding to one of Hercules' mythological labours. The theme of Hercules, a defining characteristic of Tivoli since the Republican era, due to the grand shrine dedicated to him, features prominently in the villa's overall decoration. This is partly due to clear references to Ercole d'Este, Cardinal Ippolito II's grandfather.

Raffaello da Sangallo, Fountain of Venus, 1568–1569, courtyard
The nymphaeum, leaning against the side wall of the nearby church, features in the central niche, surmounted by a frame with the bust of Constantine (fourth century), a statue of the sleeping deity, from the Roman period, throwing water from a vase into the white marble labrum decorated with strigilated bands and lion's heads (second century).

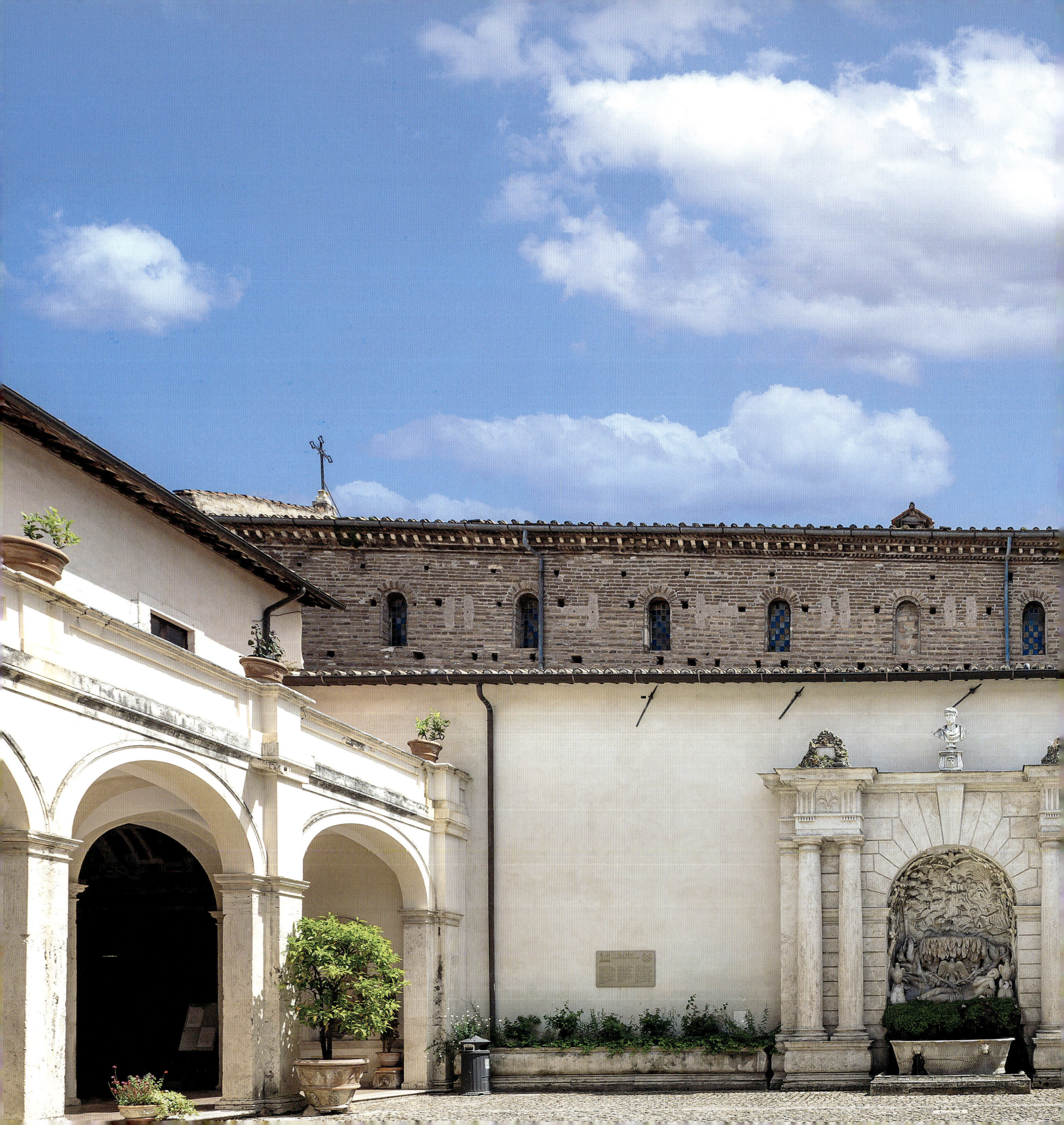

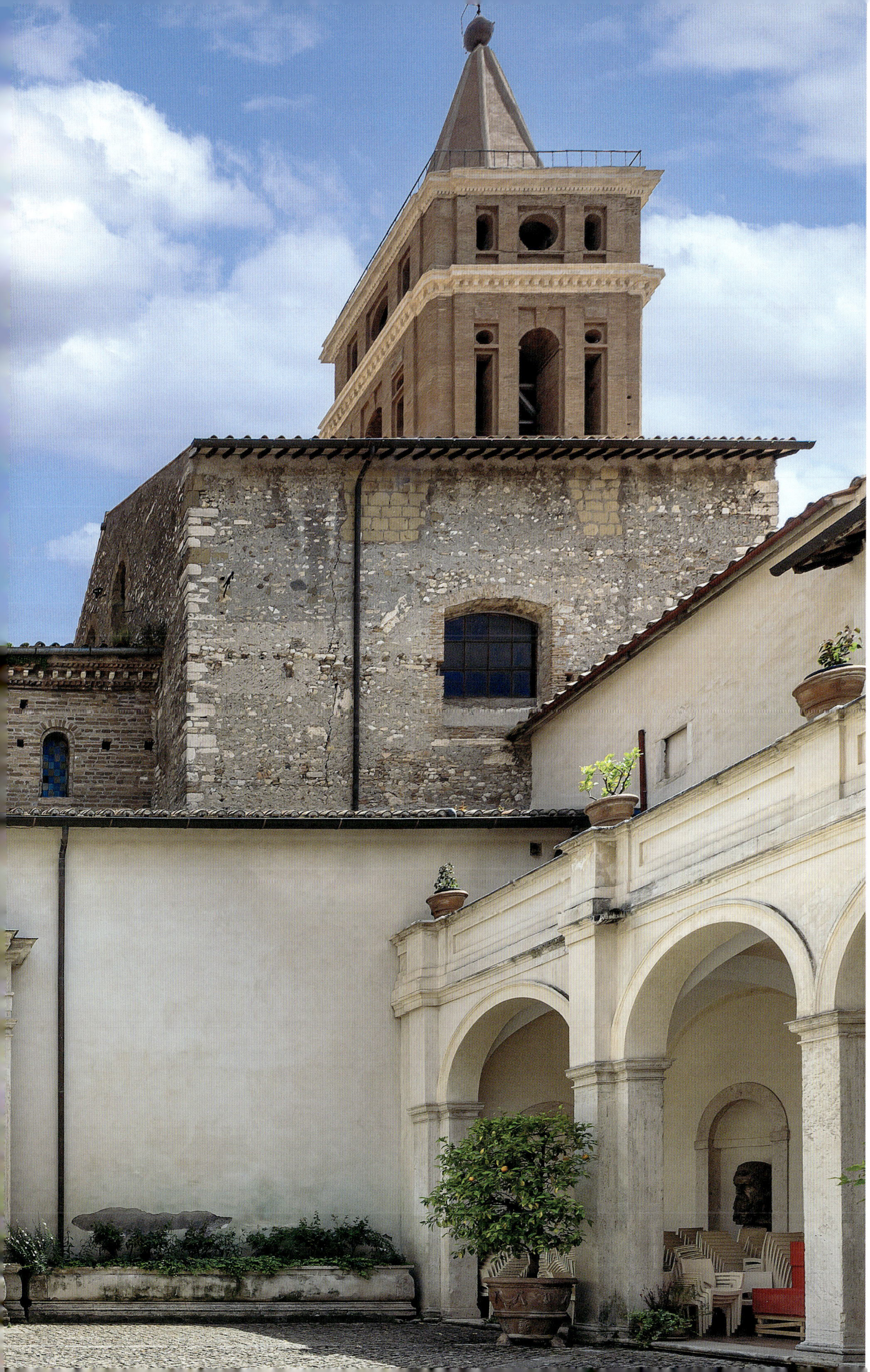

View of the courtyard towards
the church of Santa Maria Maggiore

Pages 40–41
Raffaello da Sangallo (execution),
Courtyard portico, 1566–1567
The round-arched arcades of the portico courtyard, set on Doric pillars, by the master stonemason Sangallo, fill three sides, masking the anomalies of the layout created in the area of the former convent cloister.

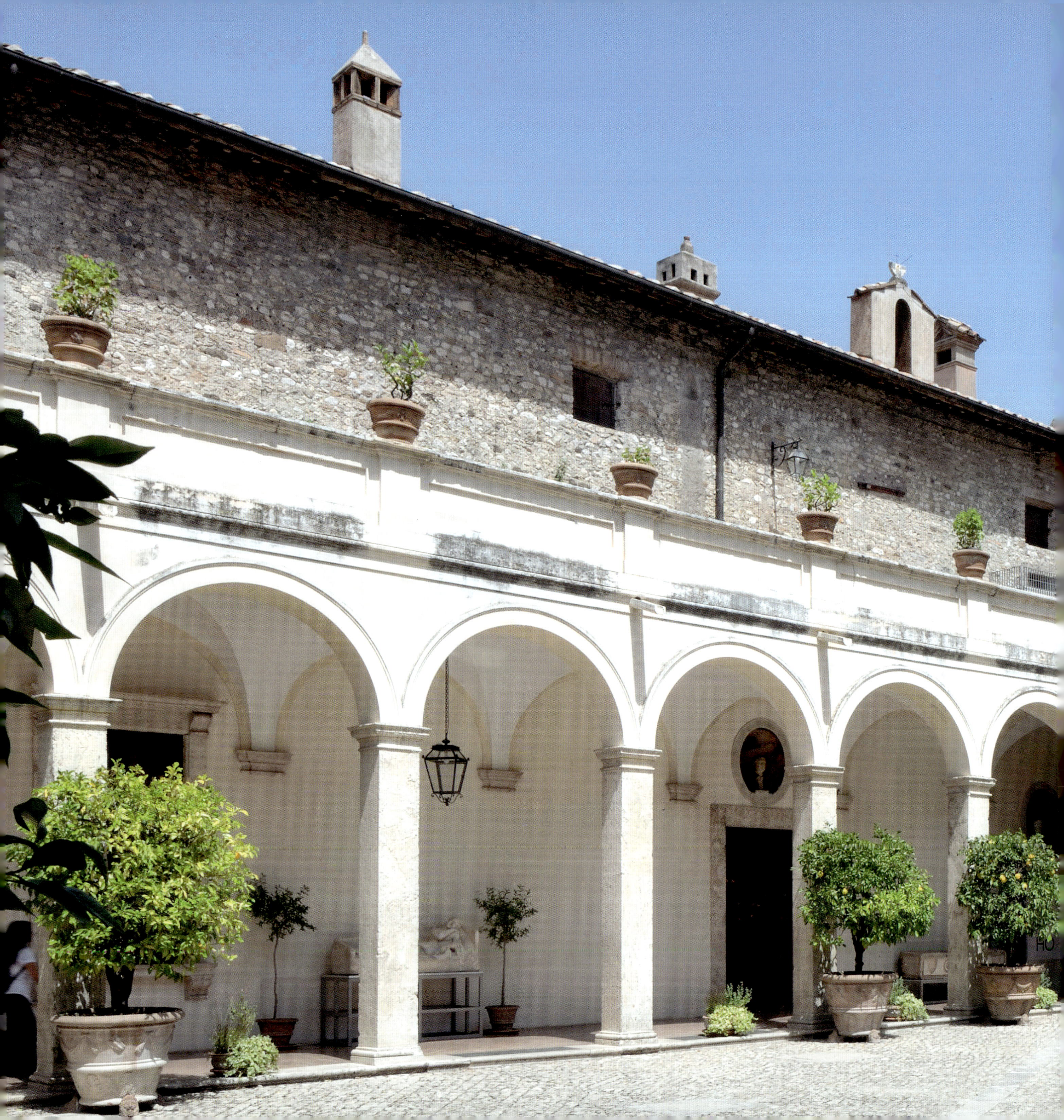

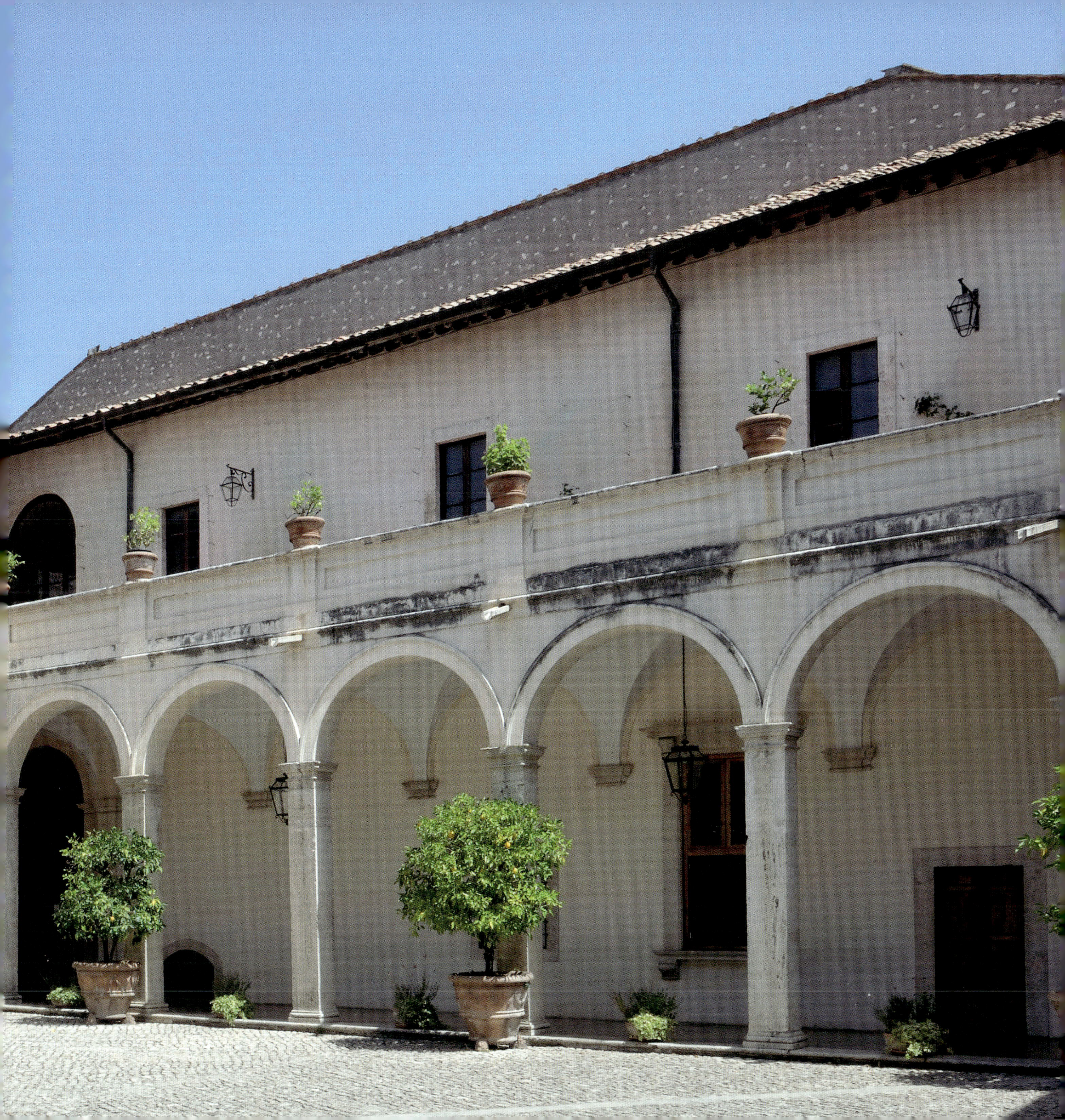

On the opposite side, there is a Roman sarcophagus lid made of Luni marble in the 'bed-shaped' style, featuring a half-reclining figure in eternal slumber, probably a sleeping Ariadne. This funerary monument was stolen from Italy and sawn into two sections for transportation but was returned in 2015, thanks to the Carabinieri for the Protection of Cultural Heritage. Reassembled and with an ancient fracture repaired, the sarcophagus was acquired by the Institute in July 2019. The inclusion of the *Ariadne* in the Villa d'Este's museographic itinerary is part of the current management's efforts to restore the villa's antiquarian character as it appeared under Cardinal Ippolito II. This also aligns with the goal of modernising the arrangement in line with contemporary sensibilities. In accordance with this approach and considering similar iconographies, the sleeping *Ariadne* is positioned in a mirrored arrangement with the sleeping *Venus* of the fountain, enabling a silent dialogue between the two sculptures.

Apartments

From the courtyard, access to the owner's apartments is possible. Originally, these apartments may have been divided into three areas, each containing several rooms. The first area, near the entrance, was largely reconstructed after damage from bombing and now serves as a ticket office and reception area. The second area, including the Sala d'angolo (Corner room), also suffered heavy war-related damage. The third area, consisting of the rooms used as Cardinal Ippolito II's residence, retains more of the wall and floor decorations as it progresses westward. The grandest room, known as the Sala delle Virtù (Hall of the Virtues), offers access to the cardinal's private chambers and retains ceiling decoration with stucco frames. Unfortunately, the original leather engravings, primarily coloured with gold (Spanish leather), have vanished. These materials arrived raw from Burgundy and Spain between 1555 and 1568 and were engraved by numerous craftsmen, whose names are no longer associated with the works today. Among the mould carvers were Mastro Giuliano and Bramante Bramanti, while Bartolomeo Roscone and the Florentine Michele di Domenico were the *orpellai* (printers/painters).

The vault decorations, acting as visual focal points, feature twenty scenes representing the Virtues (from which the room takes its name), interspersed with medallions that may have held portraits of illustrious figures, potentially even members of the Este family. Cardinal Ippolito II's insignia stands out in the corners: Este eagles and French fleurs-de-lys within white stucco ovals. This room was painted between 1568 and 1569 by a workaday painter from Forlì, Livio Agresti, aided by seven assistants. Most probably, they all worked under the careful direction of Pirro Ligorio, with whom Agresti had previously collaborated in the Vatican. Grotesques dominate the overall decoration, an element of great appeal to Ligorio. There are also landscape views, often featuring local scenery. According to documents, the room was initially furnished with a large canopy of crimson velvet (as a result of which it is also known as the Throne Room). Chairs with precious fabrics matching the canopy color, as well as luxurious drapes, adorned the space around the windows and doors. From the only door towards the garden, it is possible to reach the external loggia, whose balustrades were originally painted golden yellow, as evidenced by their depictions visible inside the hall, in the window parapets. Renaissance majolica tiles are embedded in the terracotta flooring of the loggia. The view from the loggia is impressive: the garden, which sinks to the foot of the palace, is the intimate subject of a view extending to the farthest horizon, embracing the portion of the *Ager Romanus* (the rural area around Rome) including Montecelio and Monterotondo and Rome itself, and allowing a glimpse of the southern Sabina, while on the left the historic centre of Tivoli can be seen with the furthest offshoots of the Lucretili hills.

The same group of painters mentioned earlier, responsible for the decorations in the Hall of Virtues, also took charge of adorning the next room, the Anticamera (Antechamber). Here, you can observe a similar iconographic theme as in the previous room, but with the Virtues depicted in a standing position and the vault centred on Cardinal Ippolito II's coat of arms.

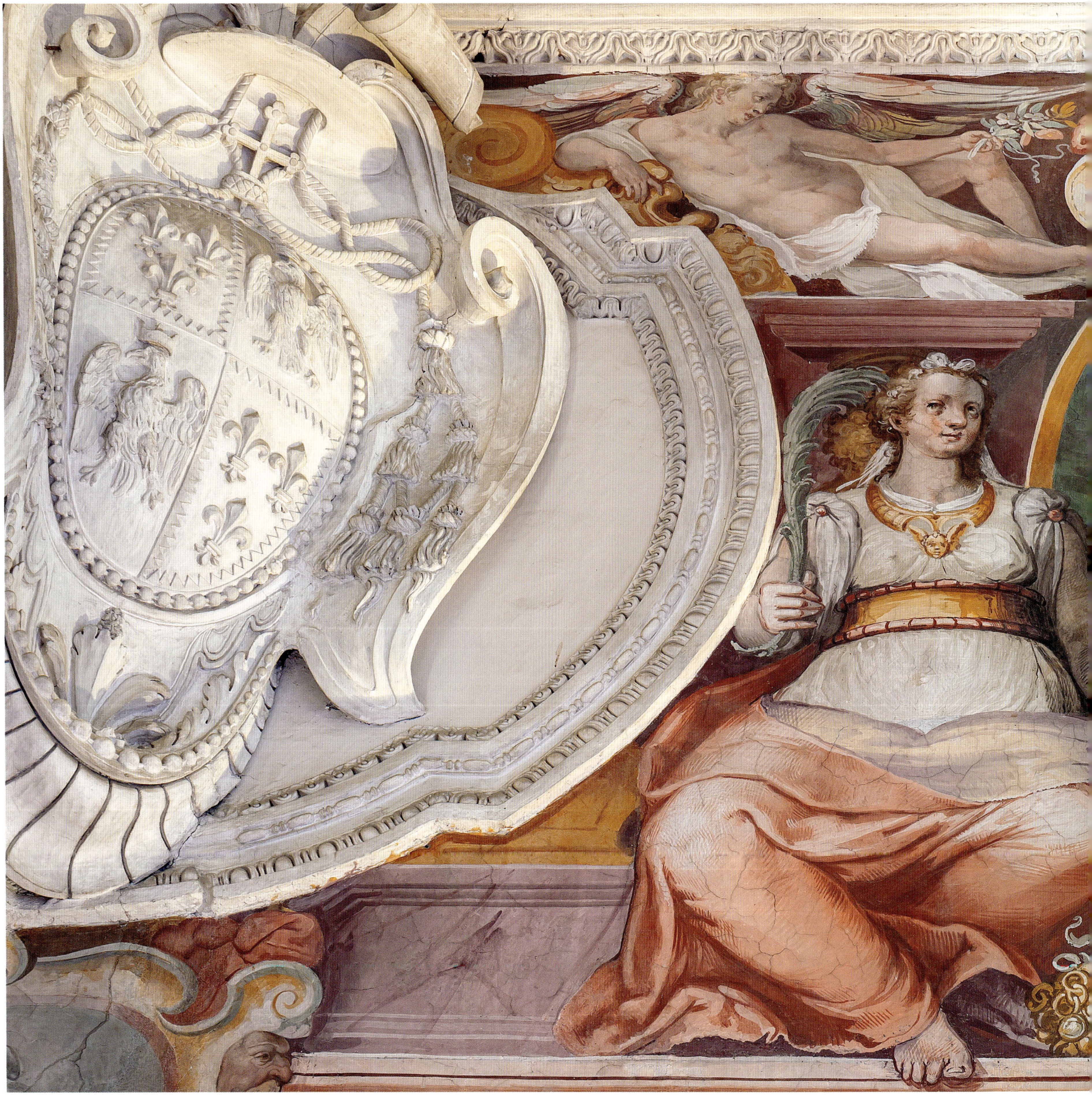

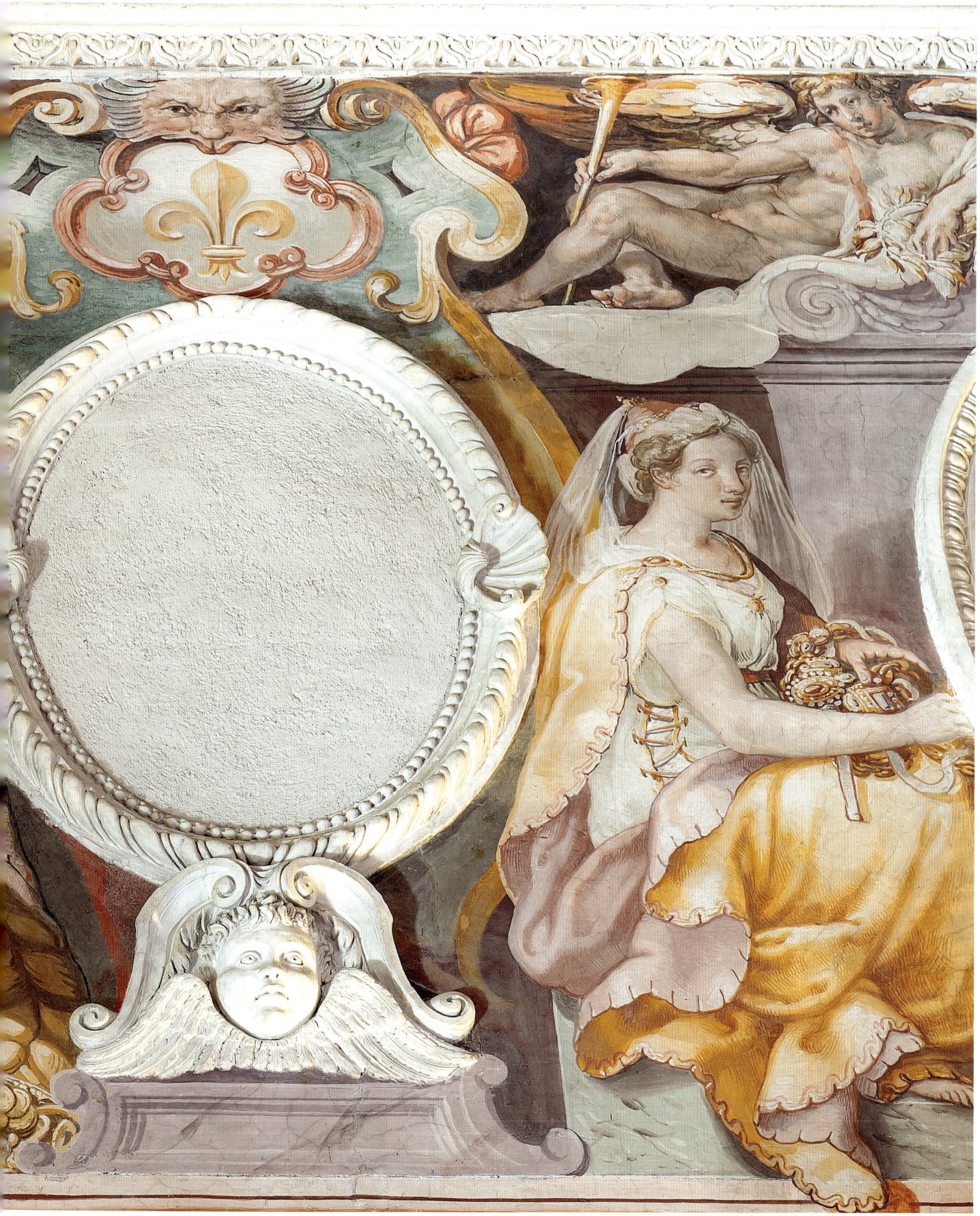

Livio Agresti, *Virtues*, 1568–1569,
Hall of the Virtues, Cardinal's Apartment

Moving on, we come to the Camera da letto (Bedroom), which boasts an exquisite wooden coffered ceiling, meticulously painted and gilded with pure gold. This masterpiece was crafted by Giovanni da Tivoli in 1569. In the centre of the ceiling, we see the cardinal's coat of arms, while the coffered panels depict scenes from Ovid's *Metamorphoses*. The cardinal's coat of arms and motto, "Ab insomni non custodita dracone" ('Not guarded by the sleepless dragon'), adorn the frieze in the subsequent room, the Sala delle Arti e dei Mestieri (Arts and Crafts Room). This room was created by Emilio Notte after extensive restoration work in the first two decades of the twentieth century. In the frieze, the figures representing various trades and crafts from the twenties give way to Renaissance Virtues.
This floor ends with the private chapel, finished before the cardinal's death in 1572. It was completed by Federico Zuccari and his assistants. The chapel features intricate and illusionistic architectural designs that foreshadow the inclusive and all-encompassing decorations of the lower floor. In the centre, above the altar, there is a fresco of the *Madonna della ghiara* (Madonna of the acorn). The adjacent room boasts a fine and original terracotta flooring. From this room it was possible to reach the *piano nobile*, which houses the noble apartments on the lower level. This access was provided through a narrow spiral staircase positioned to the side of the Manica breve (Short Corridor). The Manica breve itself is a wide corridor that, at its end, leads to the terrace over the Arco di trionfo (Triumphal Arch), offering a breathtaking view from above of the garden with the picturesque backdrop of the Gran Viale. Upon returning to the palace and following the Short Corridor, one can descend to the lower level after re-entering the courtyard, which gives access to the main staircase.

Rooms on the *piano nobile*

At the end of the grand staircase, we enter the Manica Lunga (Long Corridor): an exceptional perspective telescope, originally intended for "salubrious and pleasant walks". This corridor features three fountains and preserves the rustic mosaic cladding, a flowery

Giovanni da Tivoli, Coffered ceiling with the coat of arms of Ippolito II, 1569, Bedroom, Cardinal's Apartment
"Leandro Romanesco" and "Battista Veneziano" were its painters and gilders.

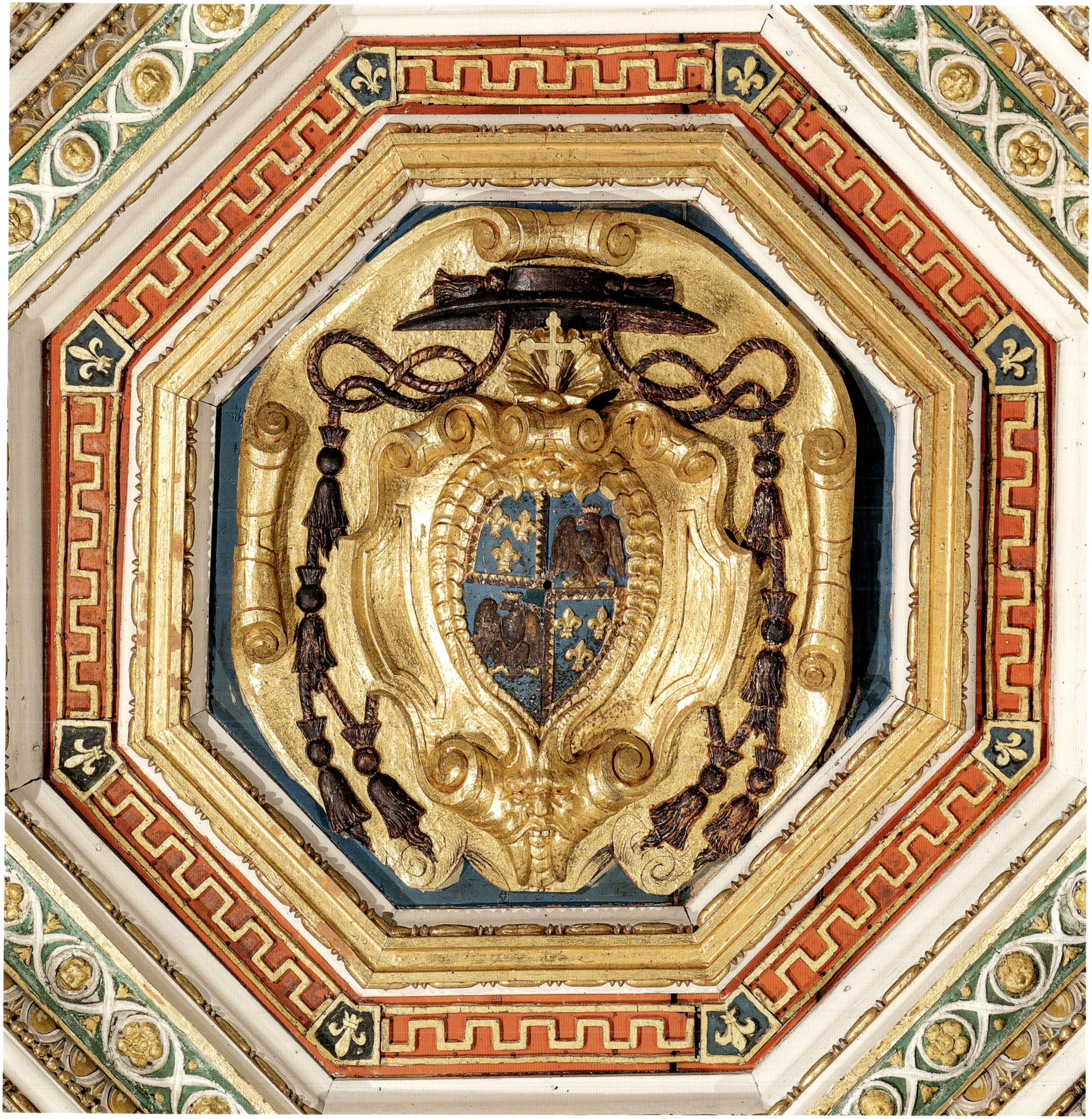

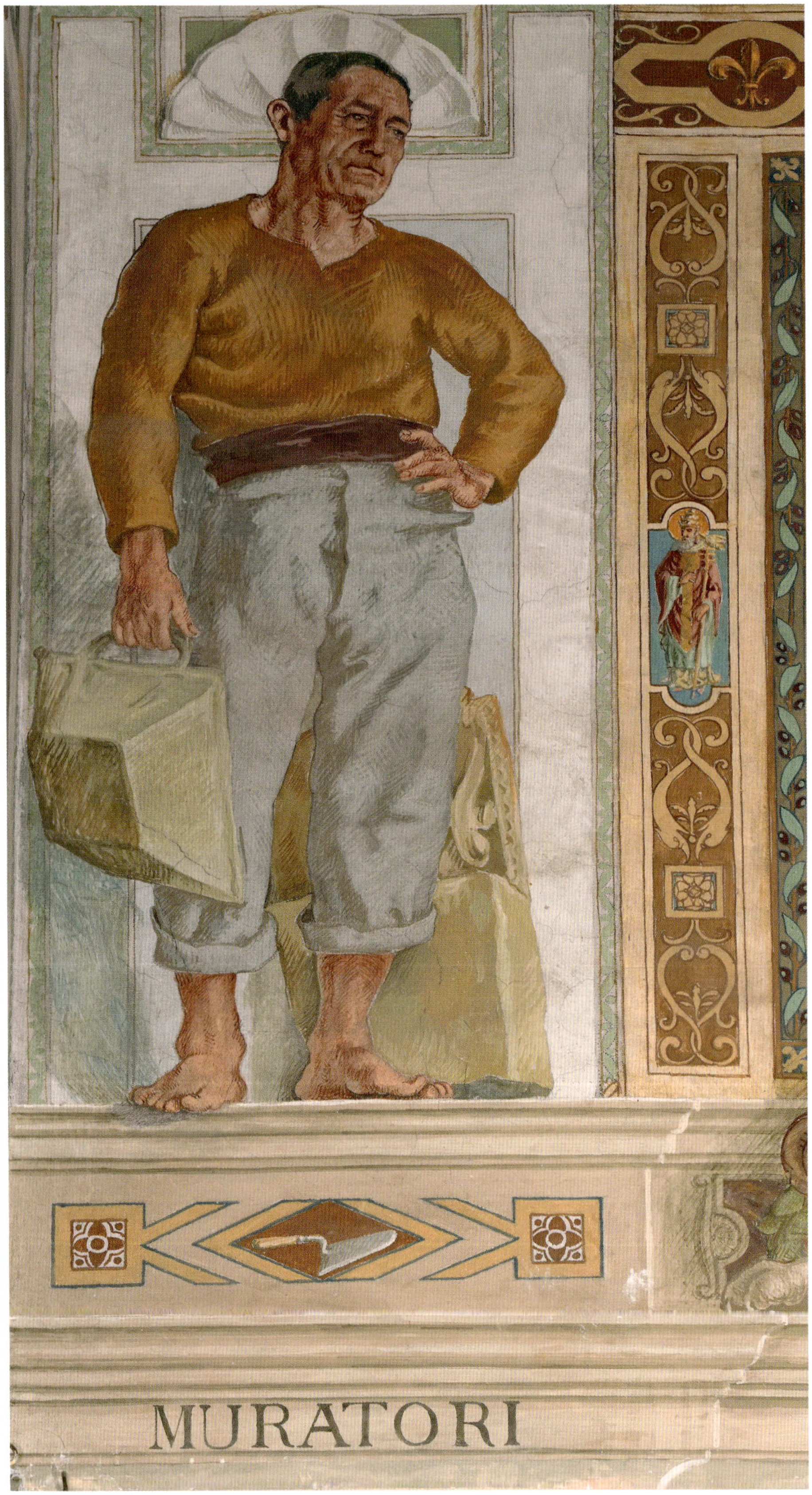

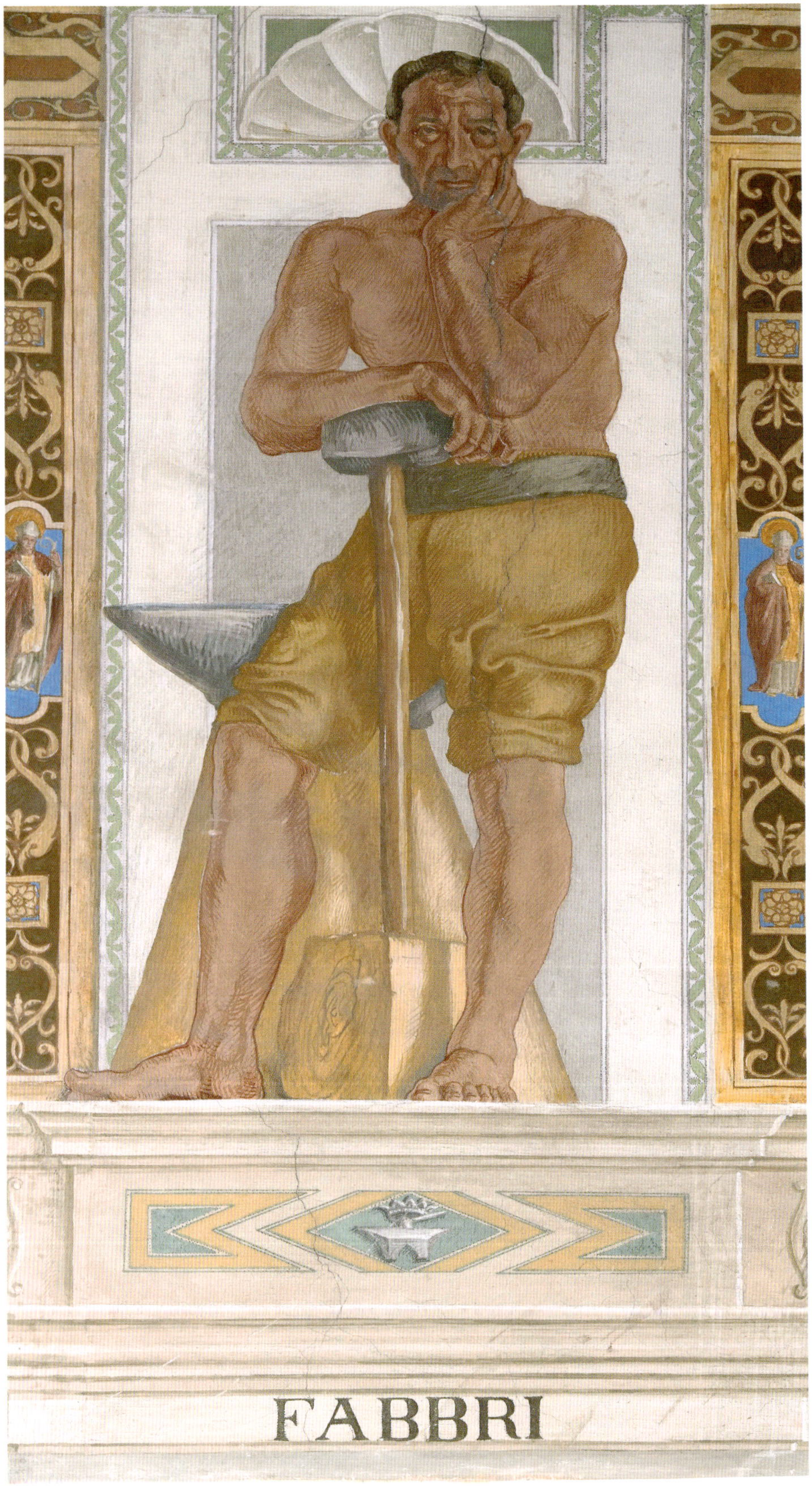

Emilio Notte, *Allegories of the Arts and Crafts*, 1920s, Arts and Crafts Room, Cardinal's Apartment

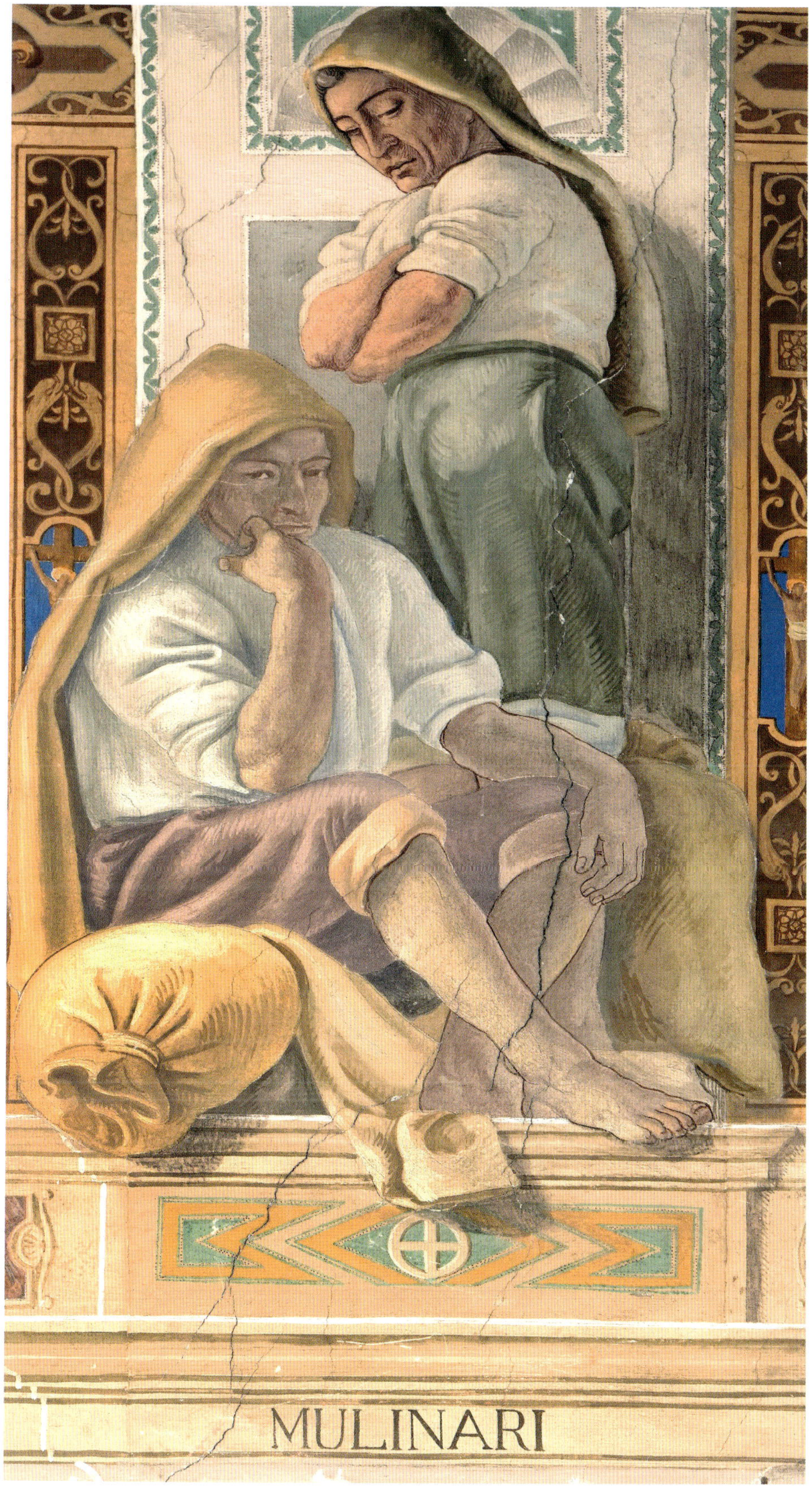
MULINARI

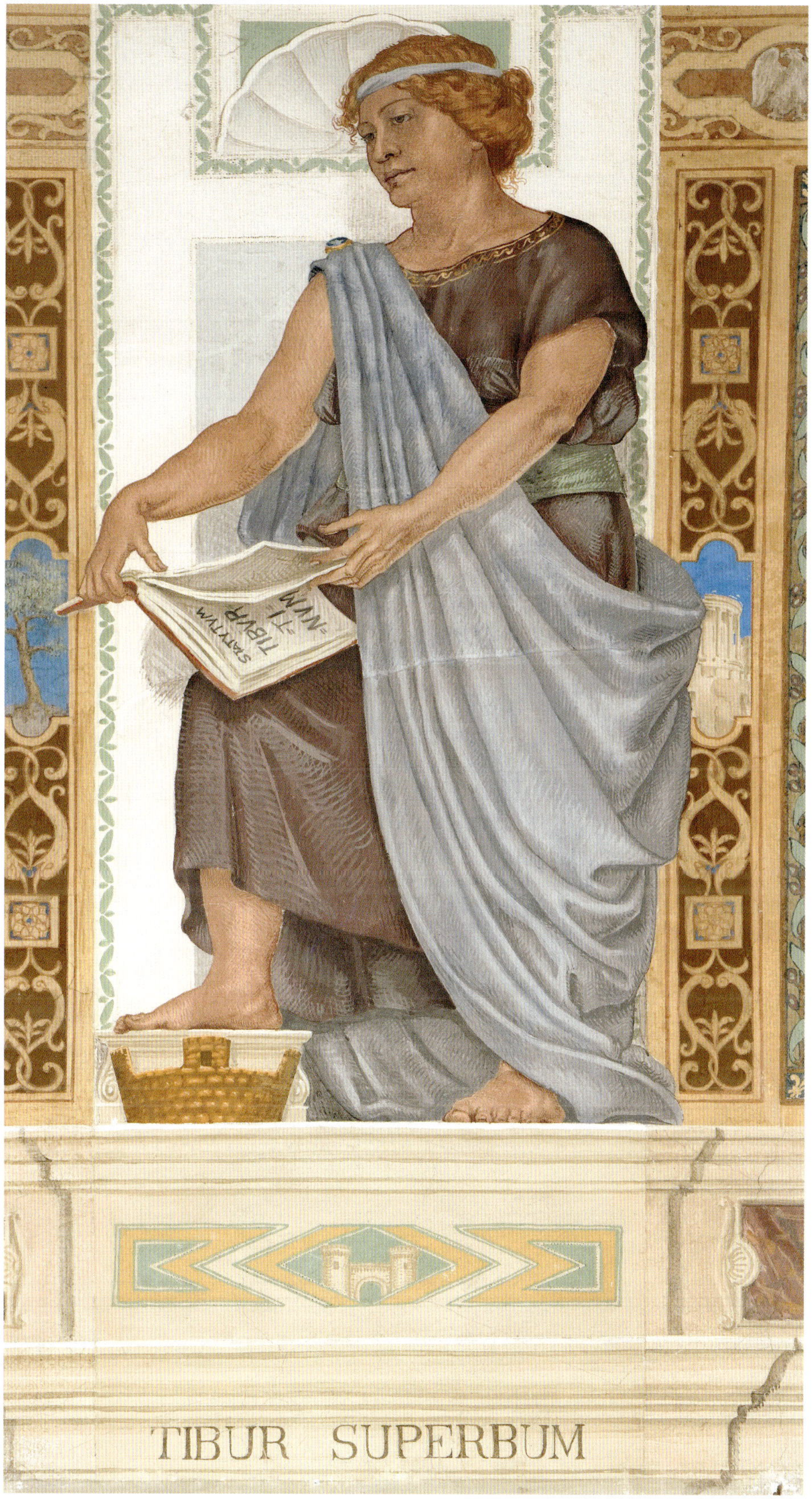
TIBUR SUPERBUM

pergola populated by birds, created with the re-use of precious marble. This space is an antiquarian reconstruction of the ancient Roman gardens, obtained by anticipating the eighteenth-century concept of the contrast between "*natura naturata*" and "*natura artificiata*", symbolising the dialogue and counterpoint between interior and exterior, between the natural environment and artificial landscape that constitutes the peculiarity of the entire Este project, creating an alternative path through nature, protected in winter and refreshing in summer heat. The construction of the pergola, which was probably intended to decorate the entire corridor, was interrupted upon the cardinal's death; in this regard, the Institute's aims have included designing a sort of vertical garden as a virtual fulfilment of Ippolito II's dream.

Turning to the right from the grand staircase, one enters the noble or summer apartment, a series of totally frescoed rooms used for receptions and banquets. The decorations have as their dominant theme the exaltation of the House of Este through its links with the myth of Hercules, the legendary progenitor of the dynasty, as well as the mythological and landscape representation of the territory, in an interchange with the local estates of which Ippolito II was governor. The first room, called Sala di Noè (Hall of Noah), features large tapestries framed in the walls, depicting Tivoli's river settings and fantastic landscapes in a profusion of detail. These are the first major examples of antiquarian landscape painting, in which the unprecedented link with the fantastic 'old Tivoli', Hadrian's residence that Ligorio was 'hollowing out', is highlighted. Rustic buildings, ruins, trees, lakes, with water and nature dominating, are illustrated in a new and unprecedented way, likely influenced by Venetian and lagoon-born painters such as Girolamo Muziano and Matteo Neroni. In the Hall of Noah, the theme of *Noah making a covenant with God after the Flood* is painted in the centre of the ceiling: the episode blatantly links the cardinal's policy of reclaiming and regulating the local waters, bringing him closer to the divine dominion of the waters. In the adjacent room, the fresco of *Moses quenching the thirst of the Hebrews fleeing from*

View of the Hall of Noah, *piano nobile*

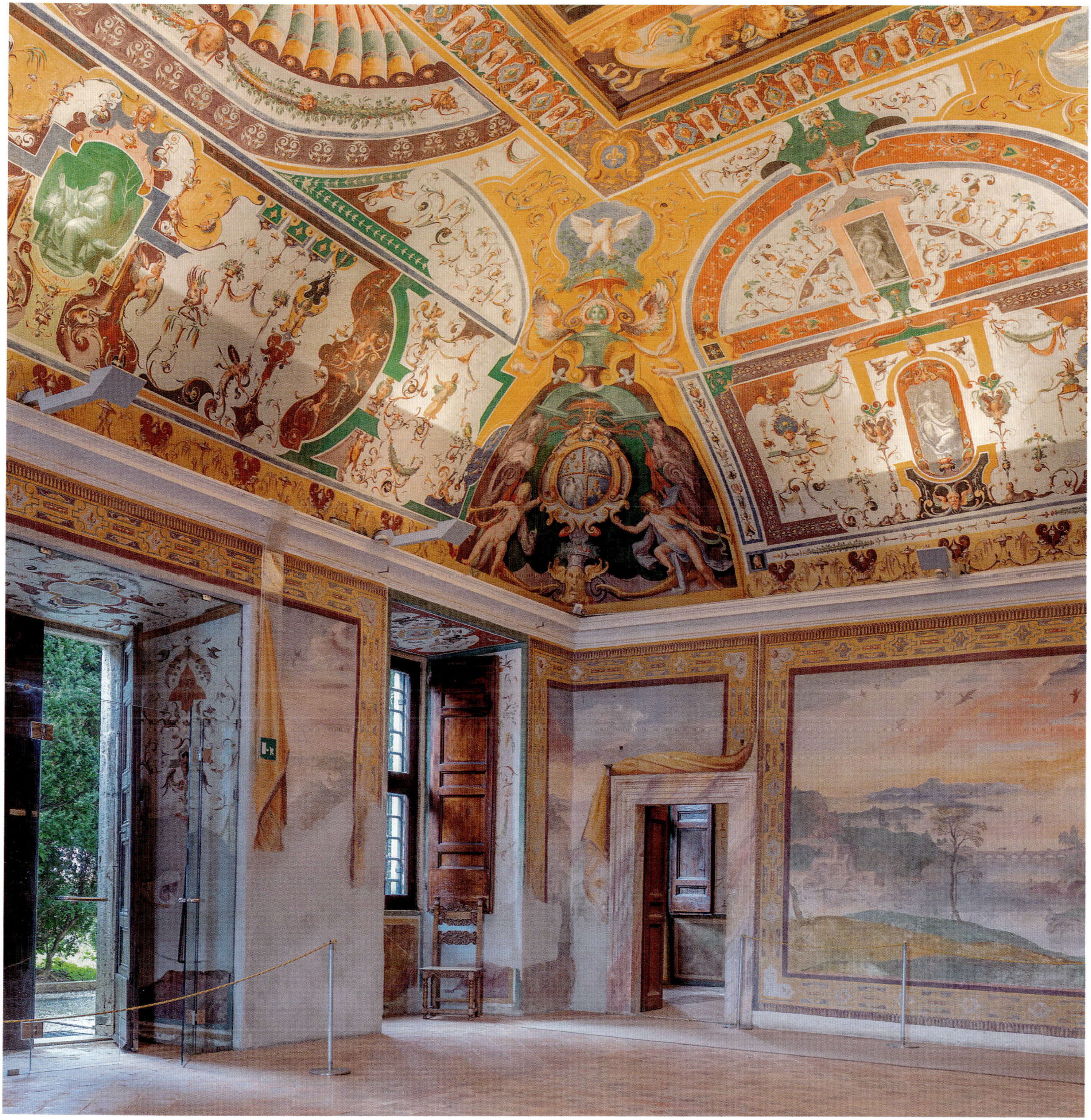

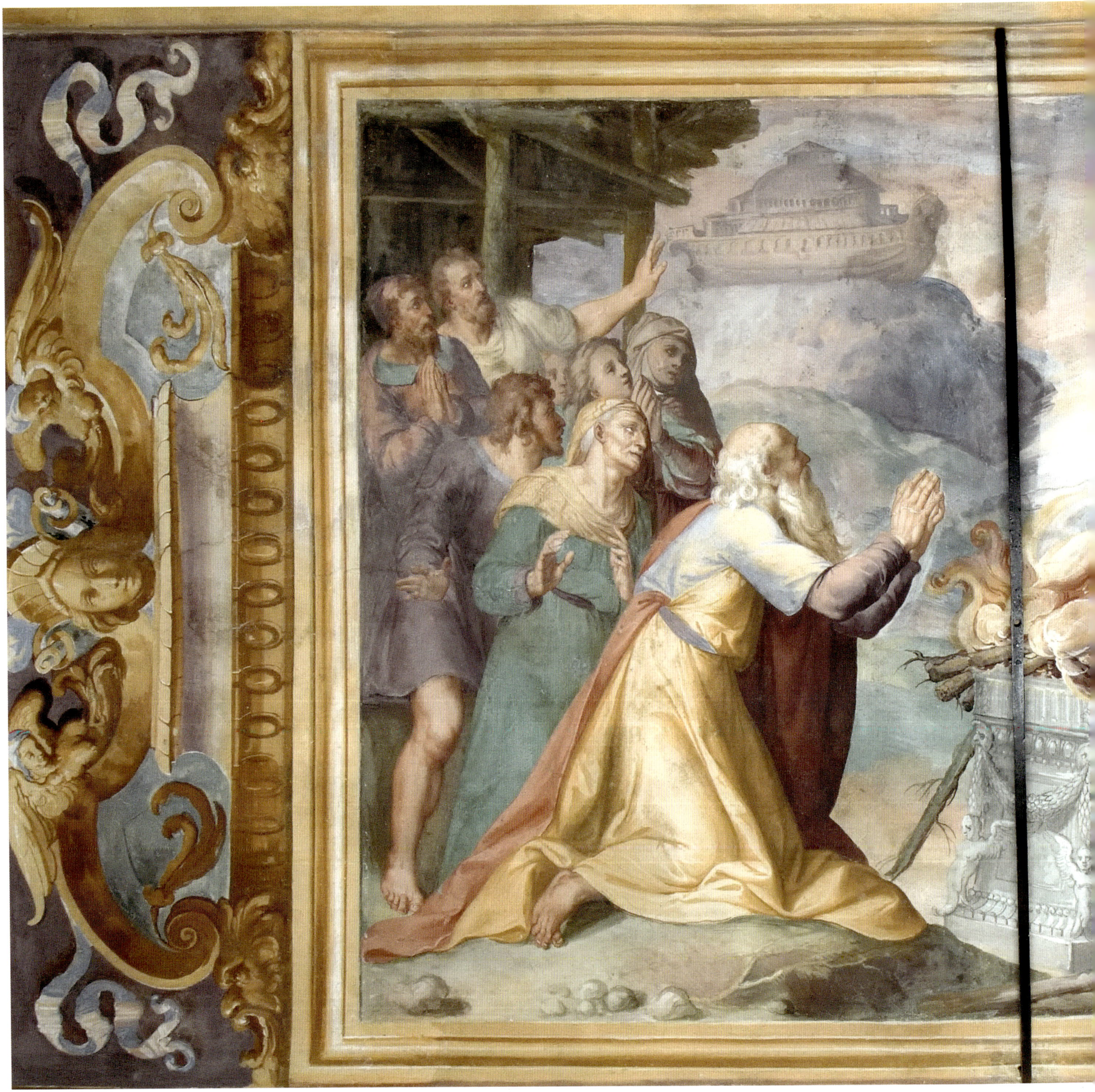

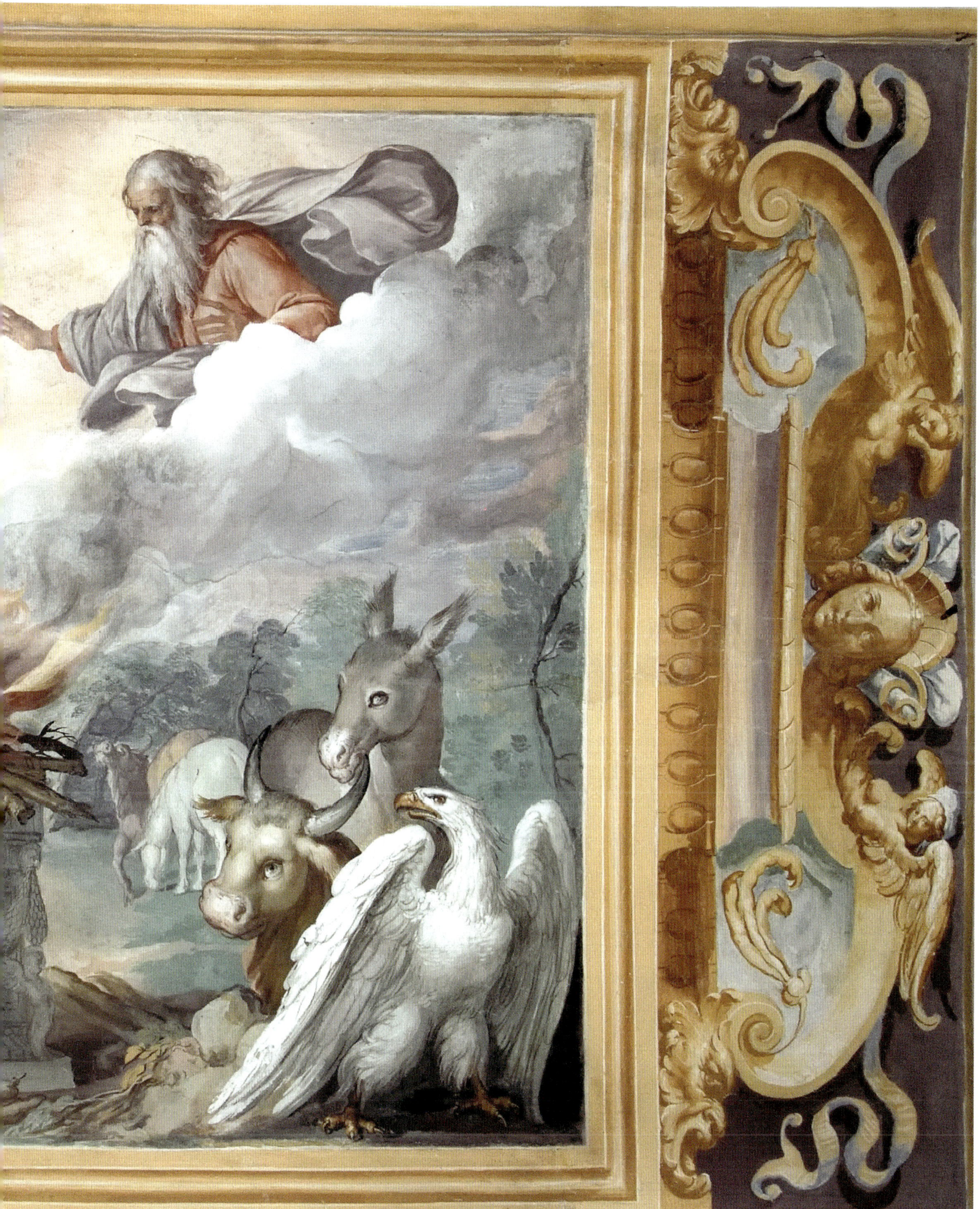

Durante Alberti, *Noah making a covenant with God after the Flood*, 1571, Hall of Noah, *piano nobile*

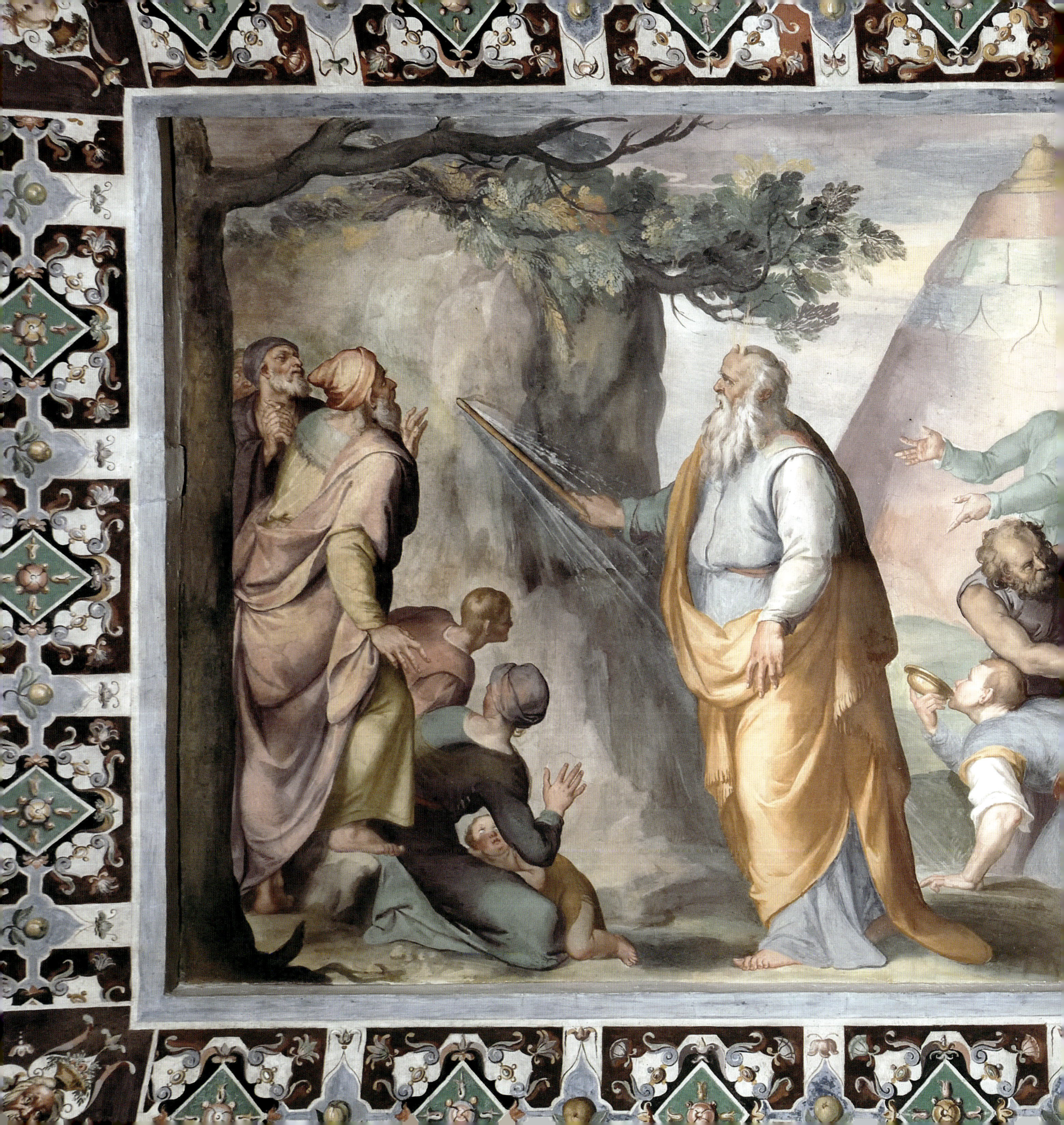

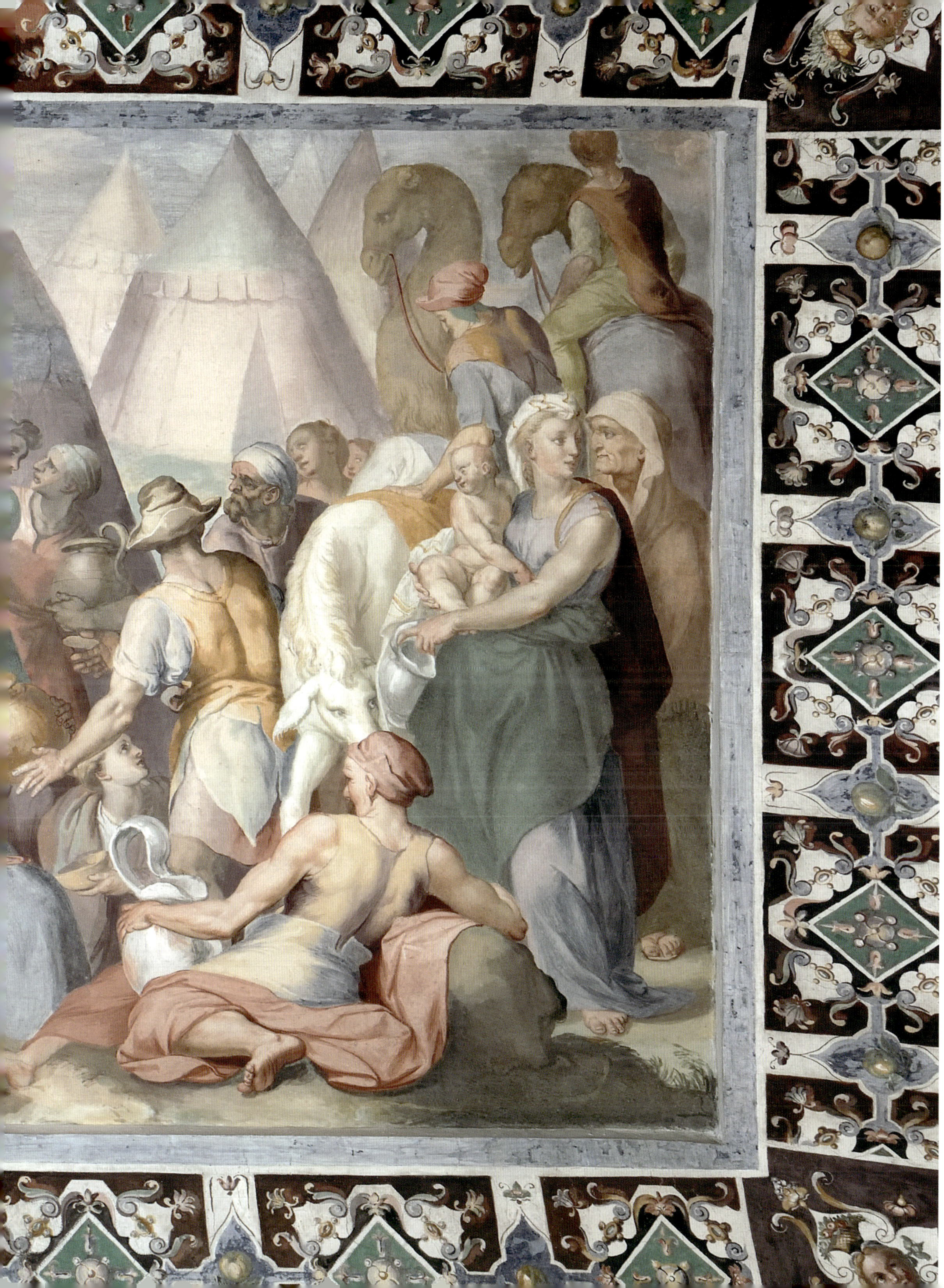

Durante Alberti and workshop, *Moses quenching the thirst of the Hebrews fleeing from Egypt*, second half of the sixteenth century, Hall of Moses, *piano nobile*

Pages 56–57
Durante Alberti and workshop, *Monochrome herms on the sides of a landscape*, second half of the sixteenth century, Hall of Moses, *piano nobile*

The room was decorated by artists directed by Durante Alberti, almost certainly under the supervision of Girolamo Muziano.

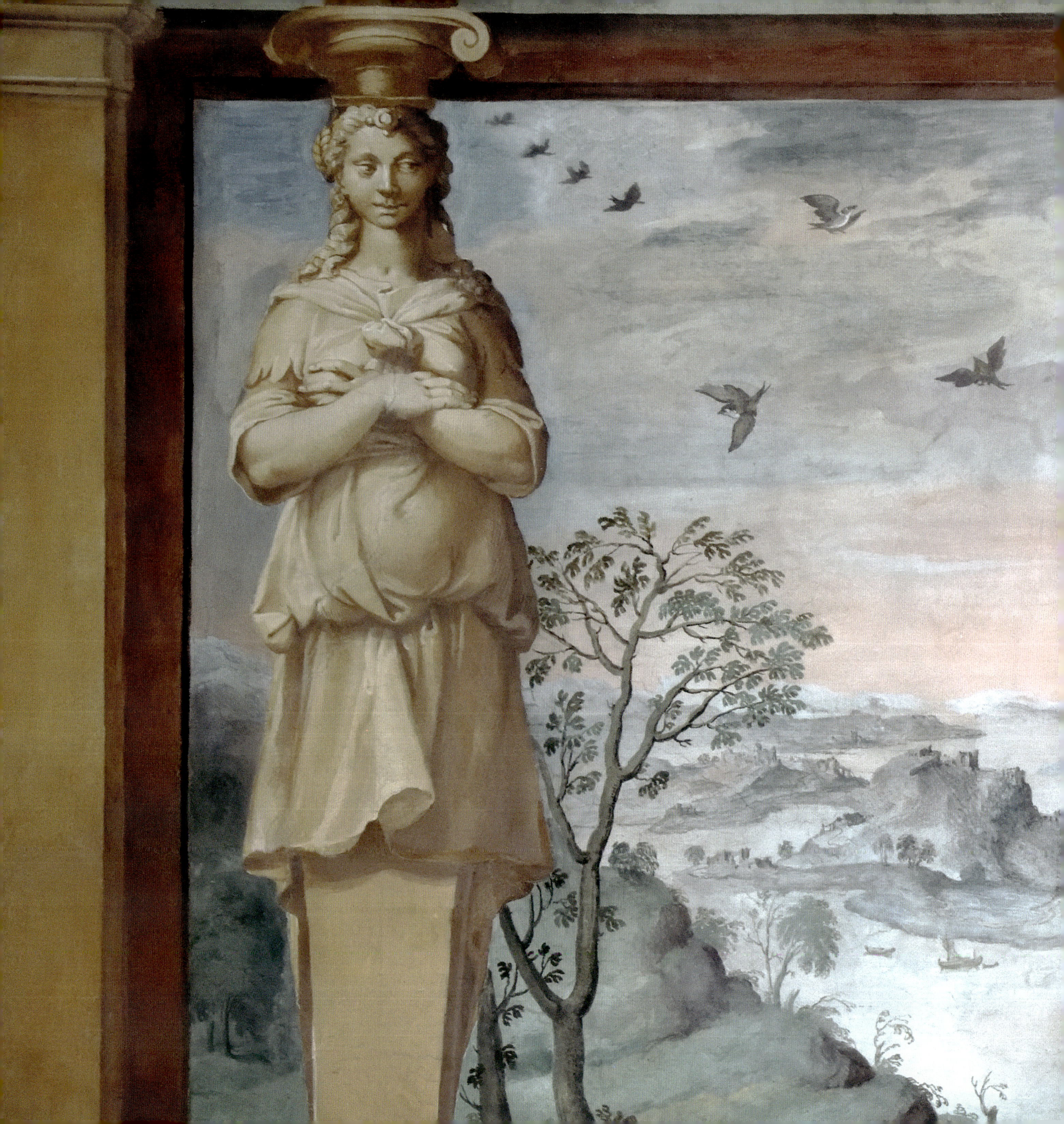

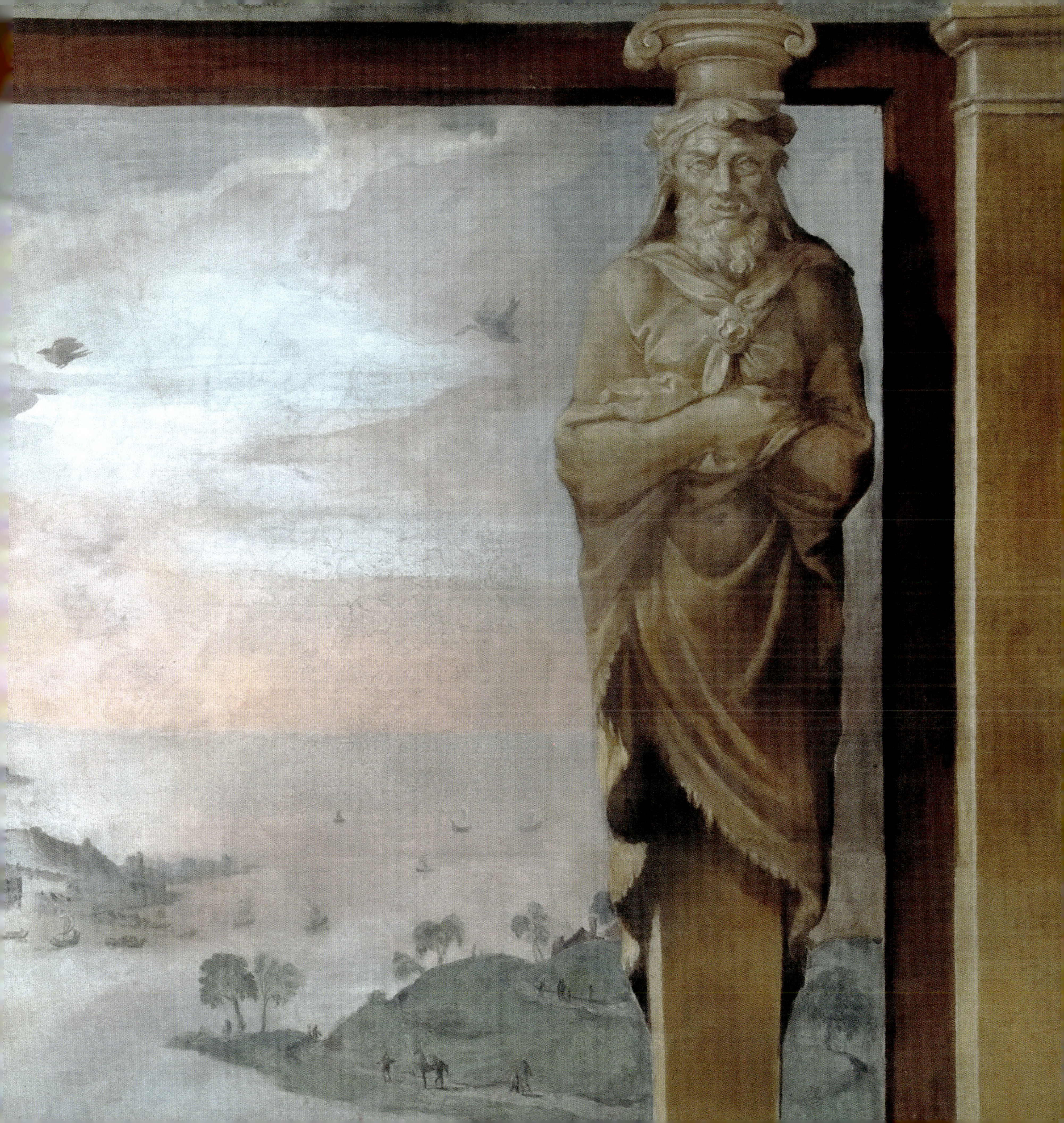

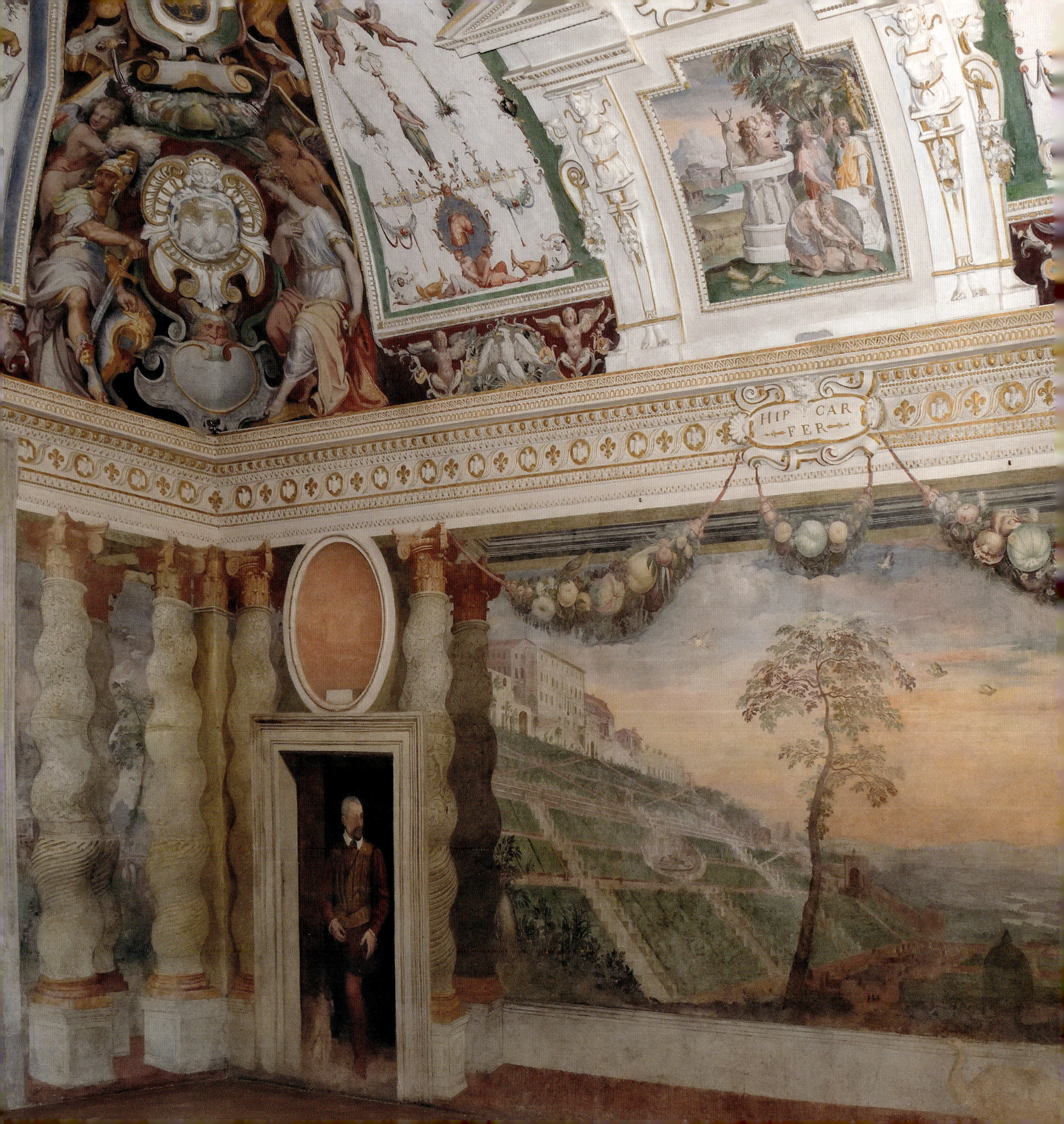
HIP CAR
FER

Egypt reaffirms the metaphor of the power of Ippolito II, who channelled the waters of the Aniene river through the center of Tivoli. Characterised by monochrome herms (squared stone pillars with a carved head on top), probably the same ones Ligorio was finding at the site of Hadrian's villa, the emblematic exchange in the room between subjects of fantasy, the result of an unprecedented syncretism between antiquarian art and natural landscape, remains a defining feature. The effect is one of great refinement and is strengthened by the original terracotta floor. The next small room, the Sala di Venere (Hall of Venus), was unfortunately greatly altered in the nineteenth century by the Christianisation programme of the Habsburg Cardinal Hohenlohe. What survives is the floor and a seventeenth-century painting on canvas on the ceiling, while a project to restore the nymphaeum and the entire room, which was originally strongly associated with the so-called Secret Garden (an external service area of the Villa), is currently under study by the Institute.

Continuing along the main path, one can admire a succession of seven entirely frescoed rooms, created by a collaboration between artists from the Zuccari and Muziano workshops. The legendary history of Tivoli is interwoven with an iconographic program that consistently depicts the exploits of Hercules at its core. These frescoes also celebrate the magnificence of the Este villa as a worthy continuation of Hadrian's legacy. The fulcrum of the itinerary is the Salone della Fontana (Hall of the Fountain), a grand reception space and a central meeting point between the garden area and the palace. This room, presumably used as a summer triclinium (a dining room), presents a decoration centred on three elements: a descriptive representation of the cardinal's properties, including the still-under-construction Villa d'Este, a ceiling with a stretched central perspective, and a fountain area, consistent in composition with the fountains in the Long Corridor. The salon's internal walls are marked by windows framing the fountains, creating a continuous illusionistic play, emphasising the dialogue between nature and artifice. The room's decor includes half-open painted doors, carefully curated artefacts, and faux twisted columns, fostering a

Detail of the Hall of the Fountain or Hall of Tivoli, *piano nobile*

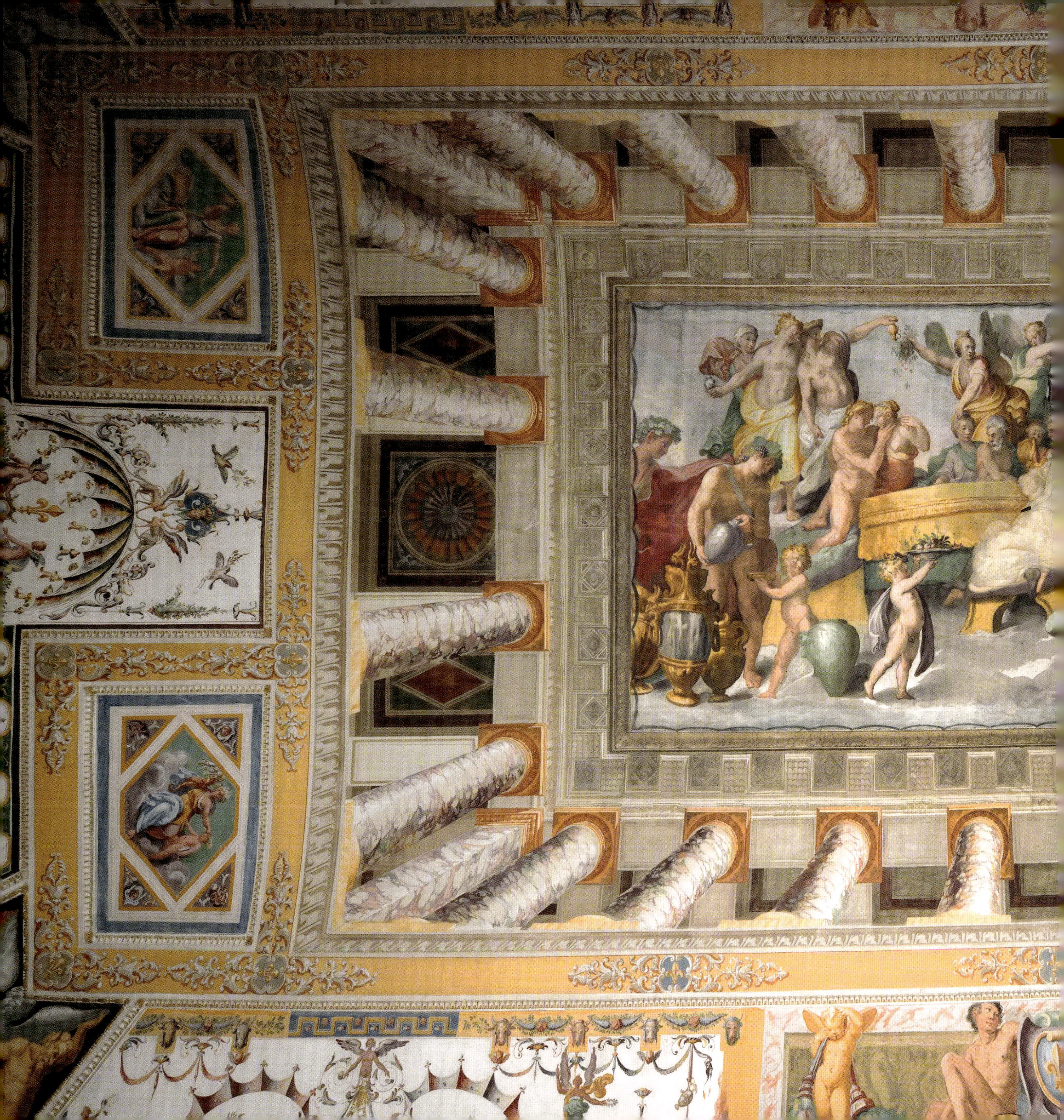

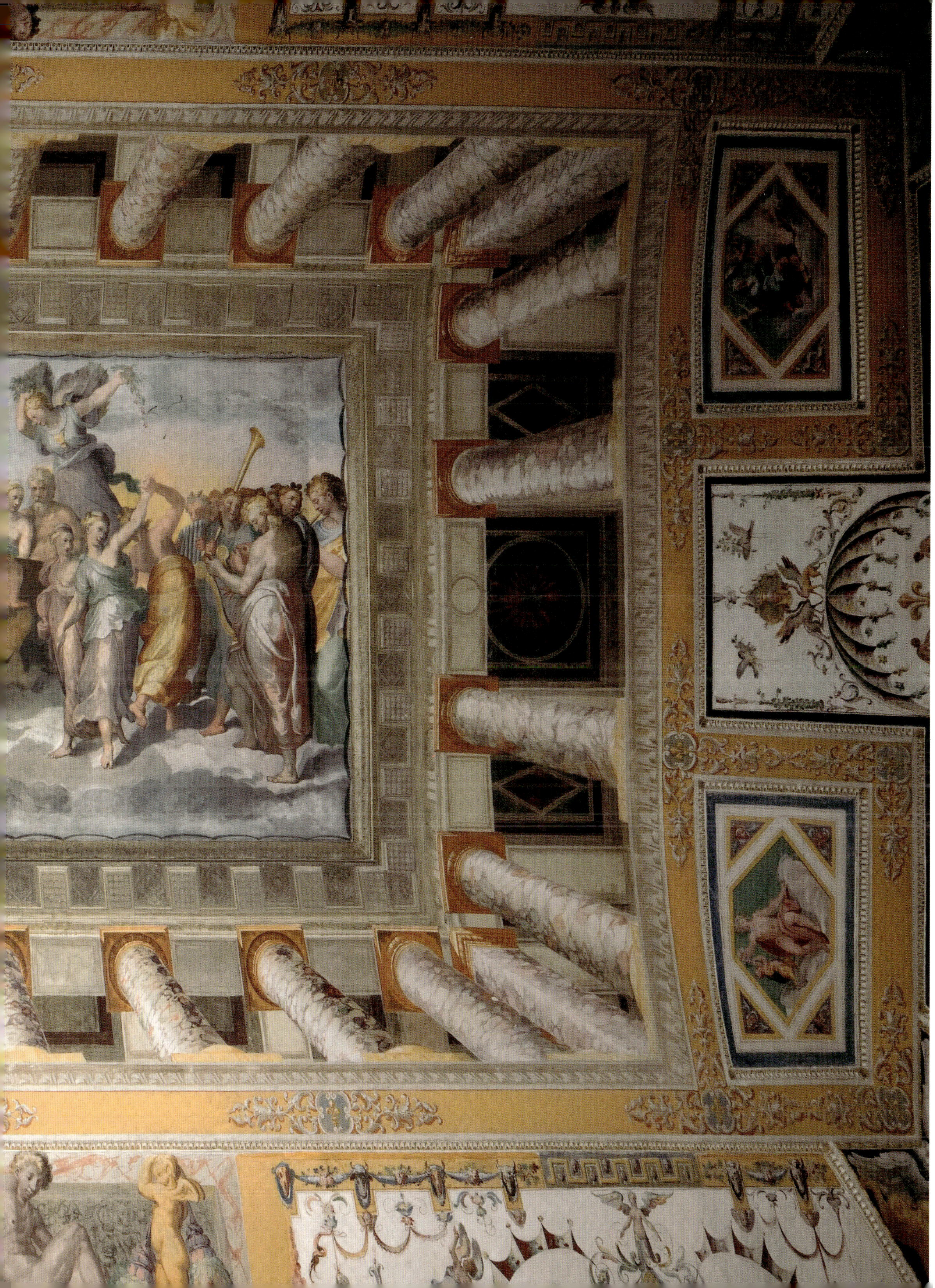

Workshop of Girolamo Muziano,
Assembly of the Gods, 1565,
Hall of the Fountain, *piano nobile*

Pages 62–63
Decoration with grotesques, 1565,
Hall of the Fountain, *piano nobile*

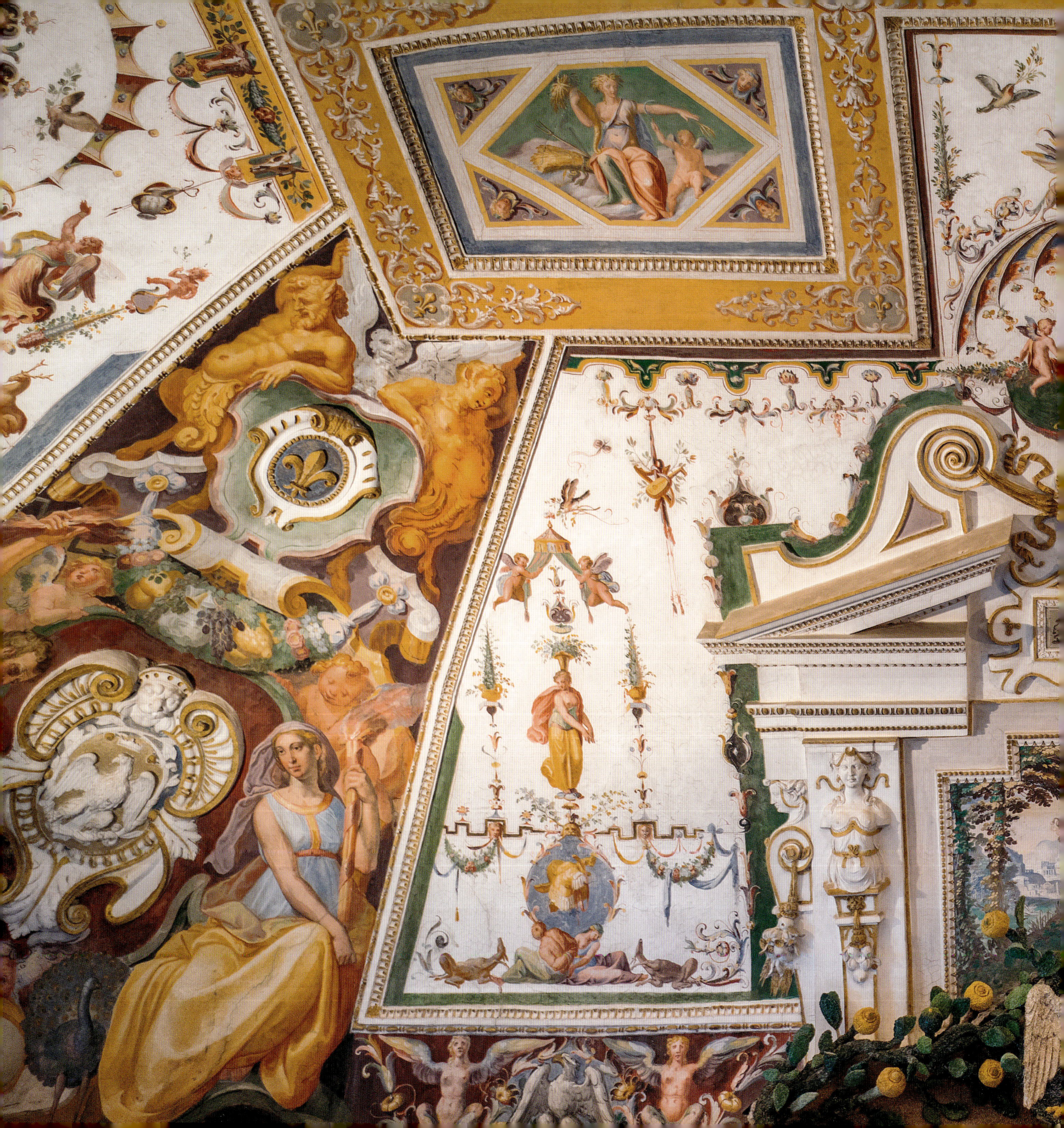

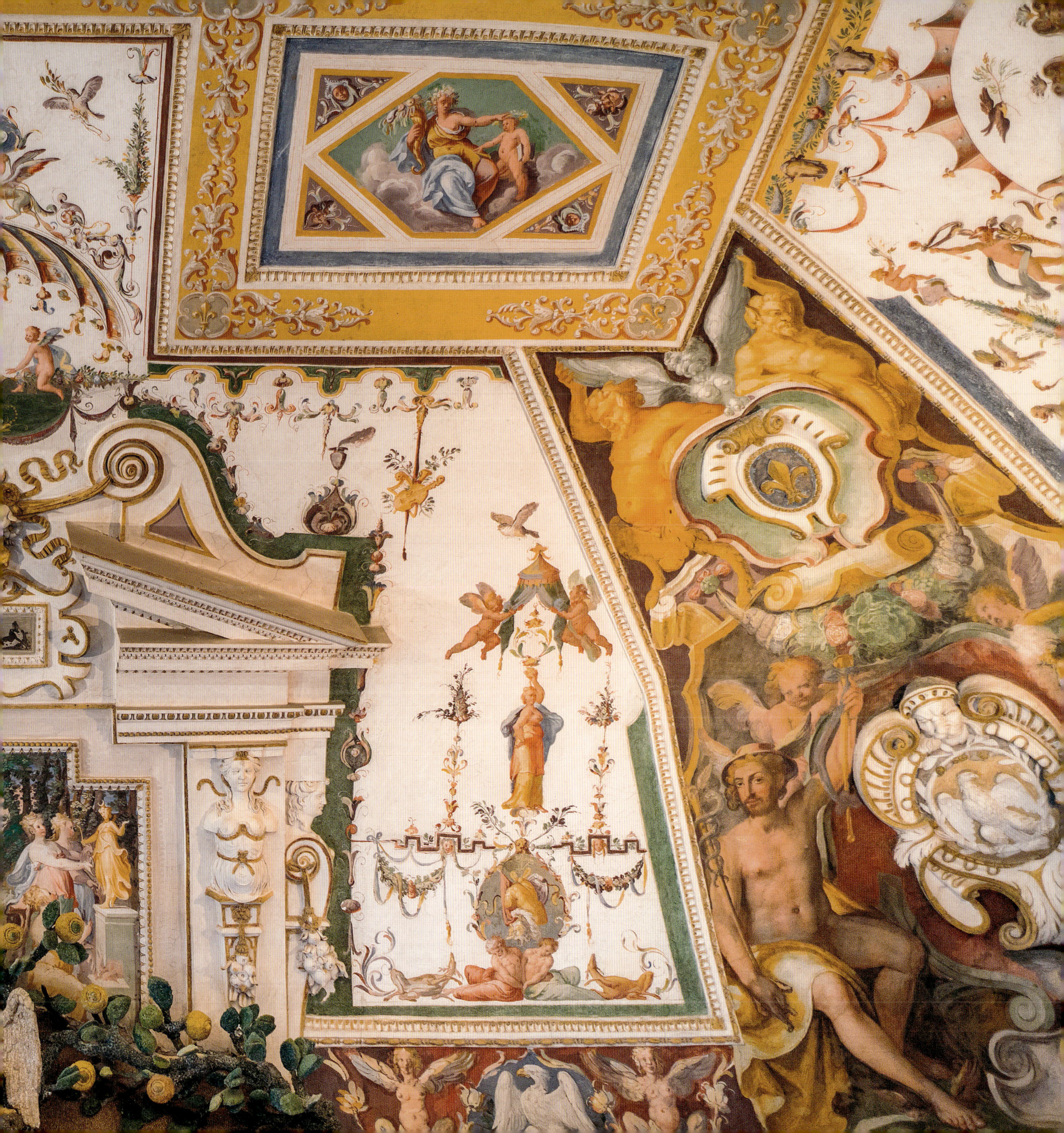

seamless exchange between the external and internal elements of the villa. Of particular interest is the depiction of the garden in its design phase and the fountain, designed by Curzio Maccarone and completed in 1568 by Paolo Calandrino, resembling a marine reconstruction of Tivoli during the Grand Tour. Likewise, the portrait of the courtly man next to the exotic primate. Next is the elegantly decorated Sala di Ercole (Hall of Hercules), possibly by Muziano, beyond which are two rooms whose paintings are from Zuccari's workshop: the Sala della nobiltà (Hall of Nobility) and the Sala della gloria (Hall of Glory). In the latter, the reference to antiquity becomes all the more evident, and the decoration – amid grotesques and Roman views – continually refers to an elegiac world studded with allegorical figures immersed in spaces that presumably evoke the mural decorations of Hadrian's Villa. These rooms are distinguished by a description characterised by great Mannerist elegance, using draperies and rich colours that pave the way for the vibrant Baroque of the next and final room, the Sala della caccia (Hunting Room), a seventeenth-century space known for its concise description of scenes with a naturalistic focus, influenced by Antonio Tempesta and Cavalier d'Arpino. In this room, game and festoons replace references to antiquity, while lake landscapes conclude the thematic connection that began with the first room – the Hall of Noah – on the *piano nobile*.

From the Hunting Room one returns to the Long Corridor, along which features connecting openings to the Terrazza della Pallacorda (Terrace of Court Tennis), now exclusively used for cultural events. The terrace, now bare, was once a court for a precursor of modern tennis, a game that interested Ippolito II as early as 1536, when he designated a space for it in his Ferrara villa at Belriguardo for the *zuogo della bala* (ball game), later also included in the villa at Fontainebleau.

It is also possible to reach the garden from the Long Corridor by descending the staircase, or, returning to the Hall of the Fountain, by walking to the loggia and descending the staircase.

Workshop of Girolamo Muziano,
Assembly of the Gods, 1565–1566,
Hall of Hercules, *piano nobile*

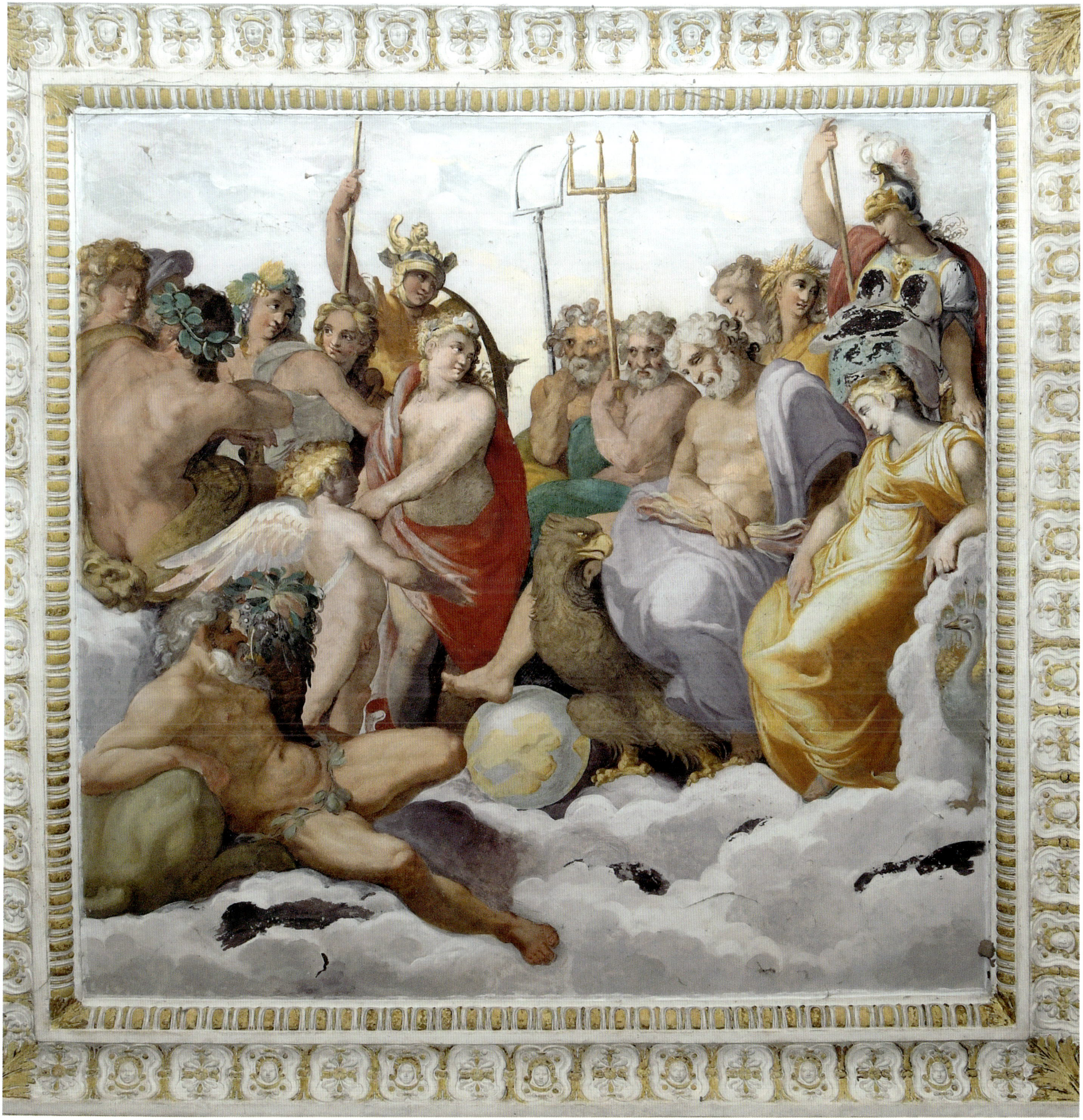

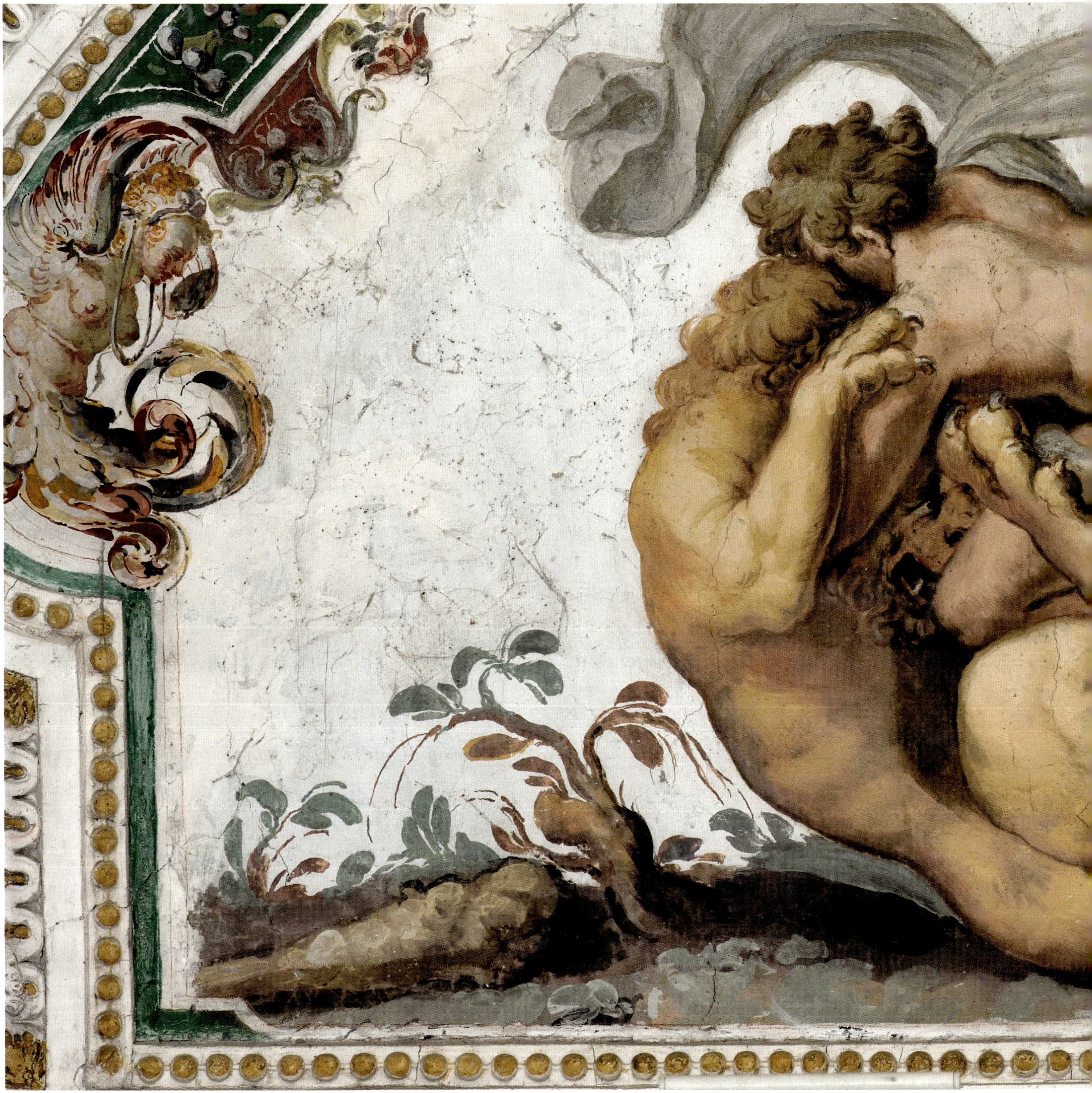

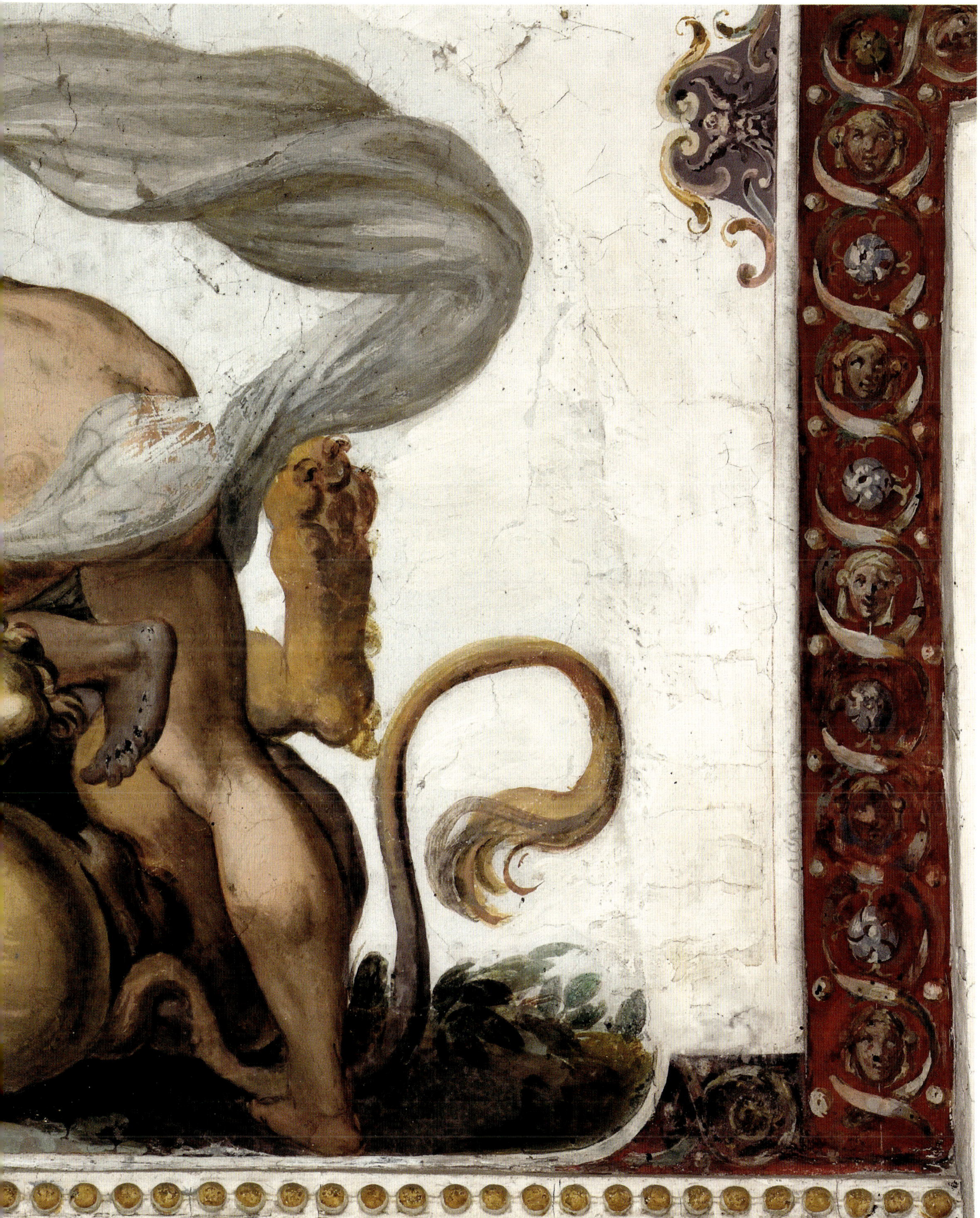

Workshop of Girolamo Muziano,
Hercules and the lion of Nemea, 1565–1566,
Hall of Hercules, *piano nobile*

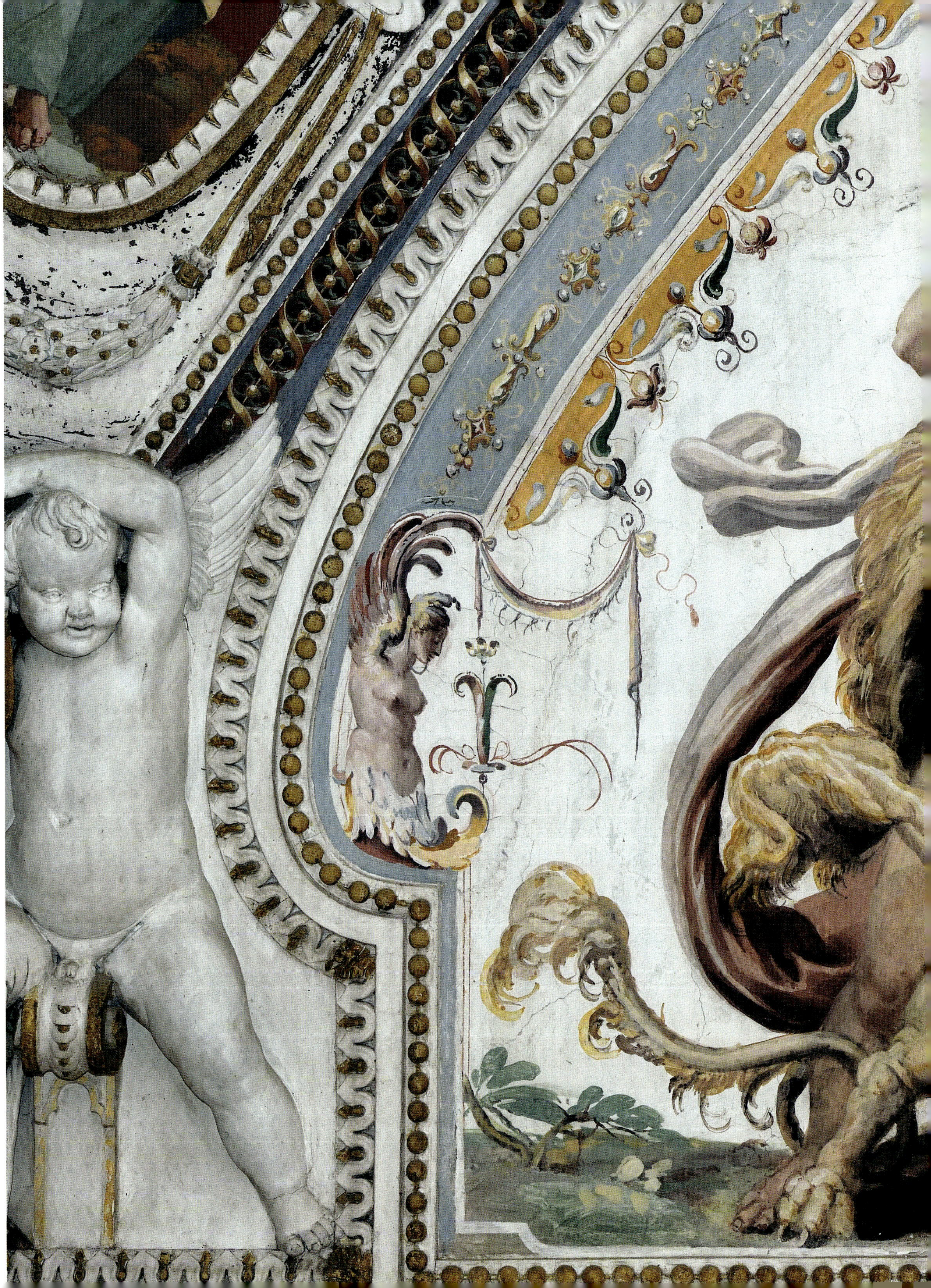

Workshop of Girolamo Muziano,
Hercules and Cerberus, 1565–1566,
Hall of Hercules, *piano nobile*

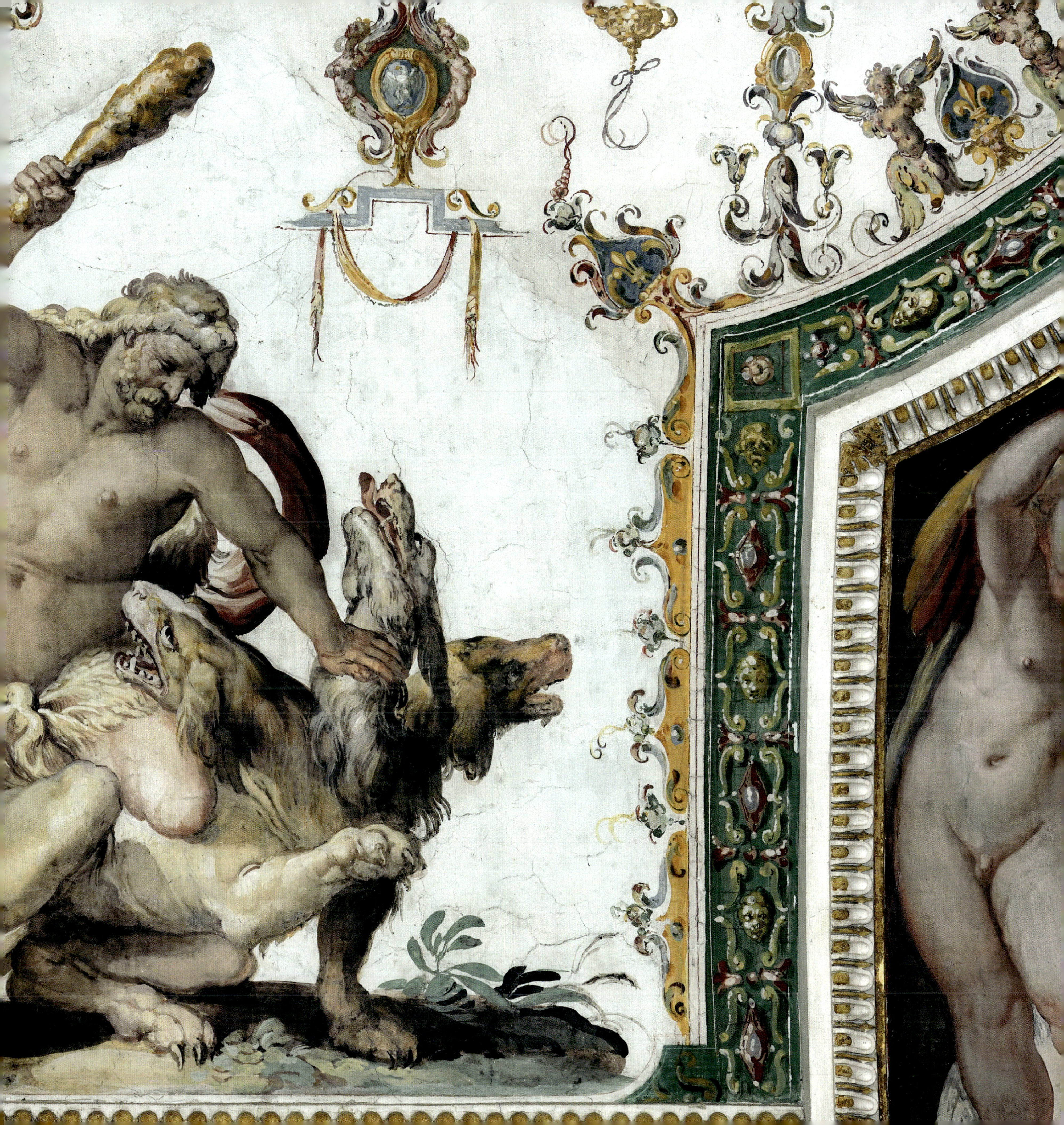

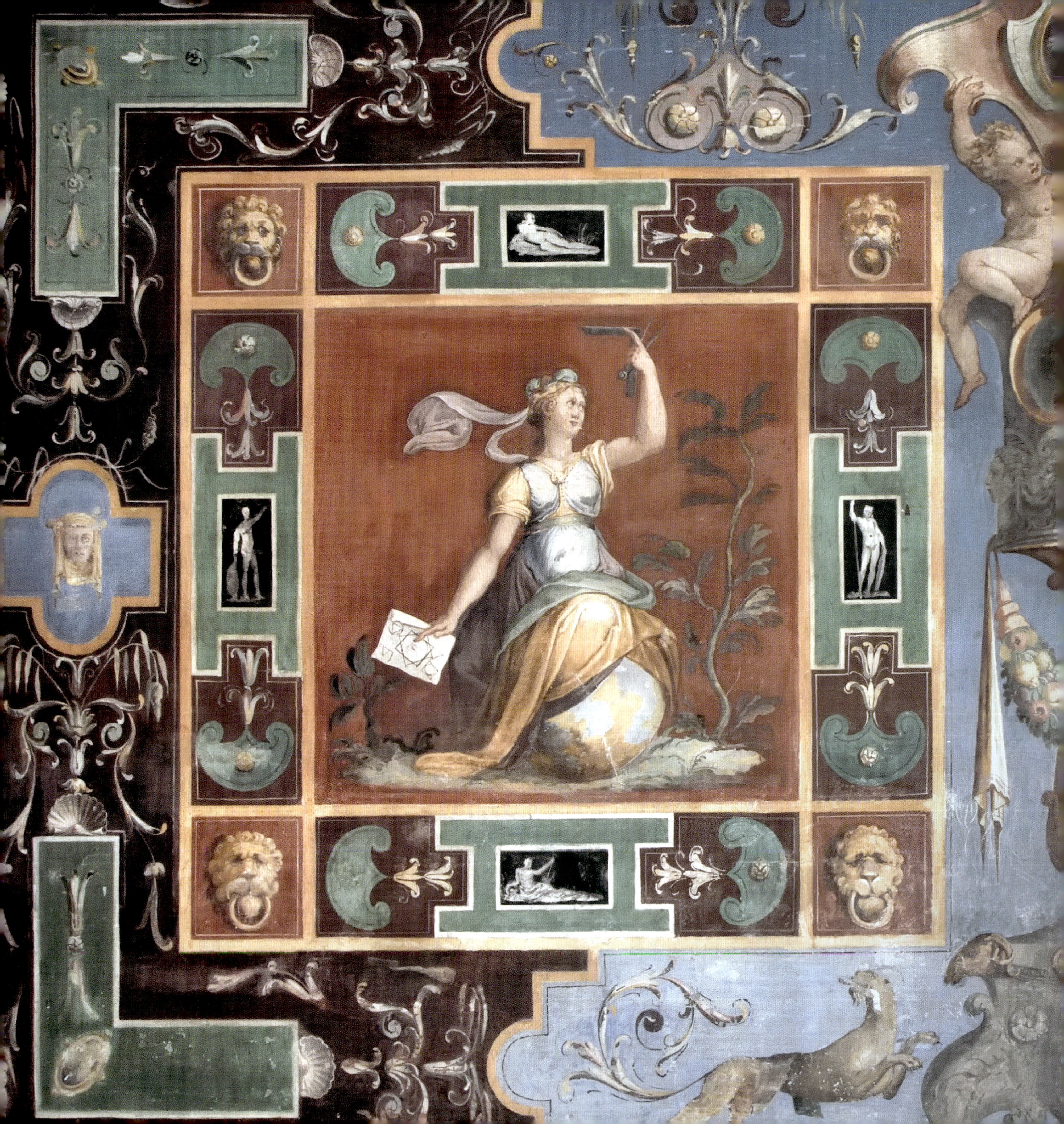

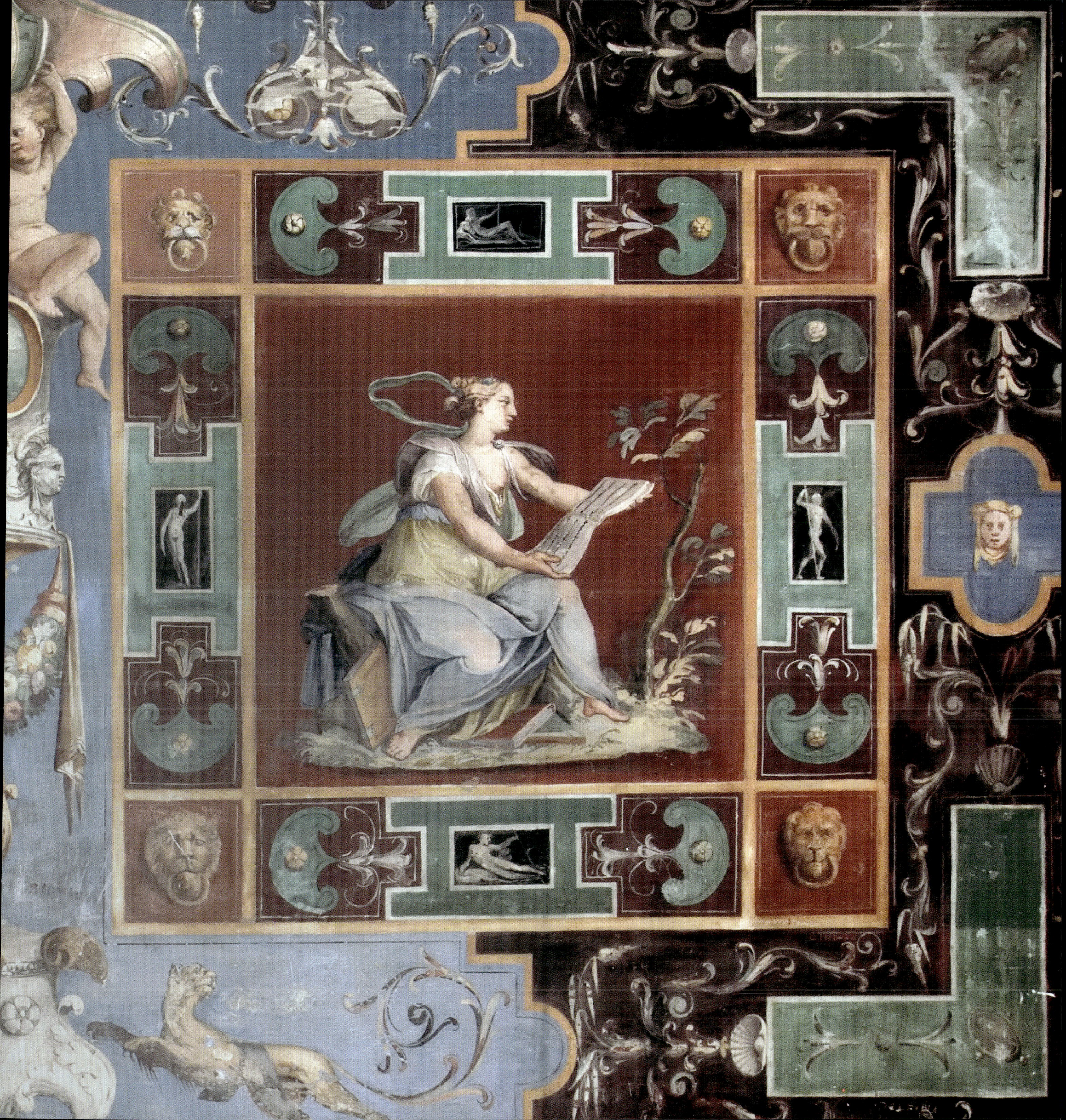

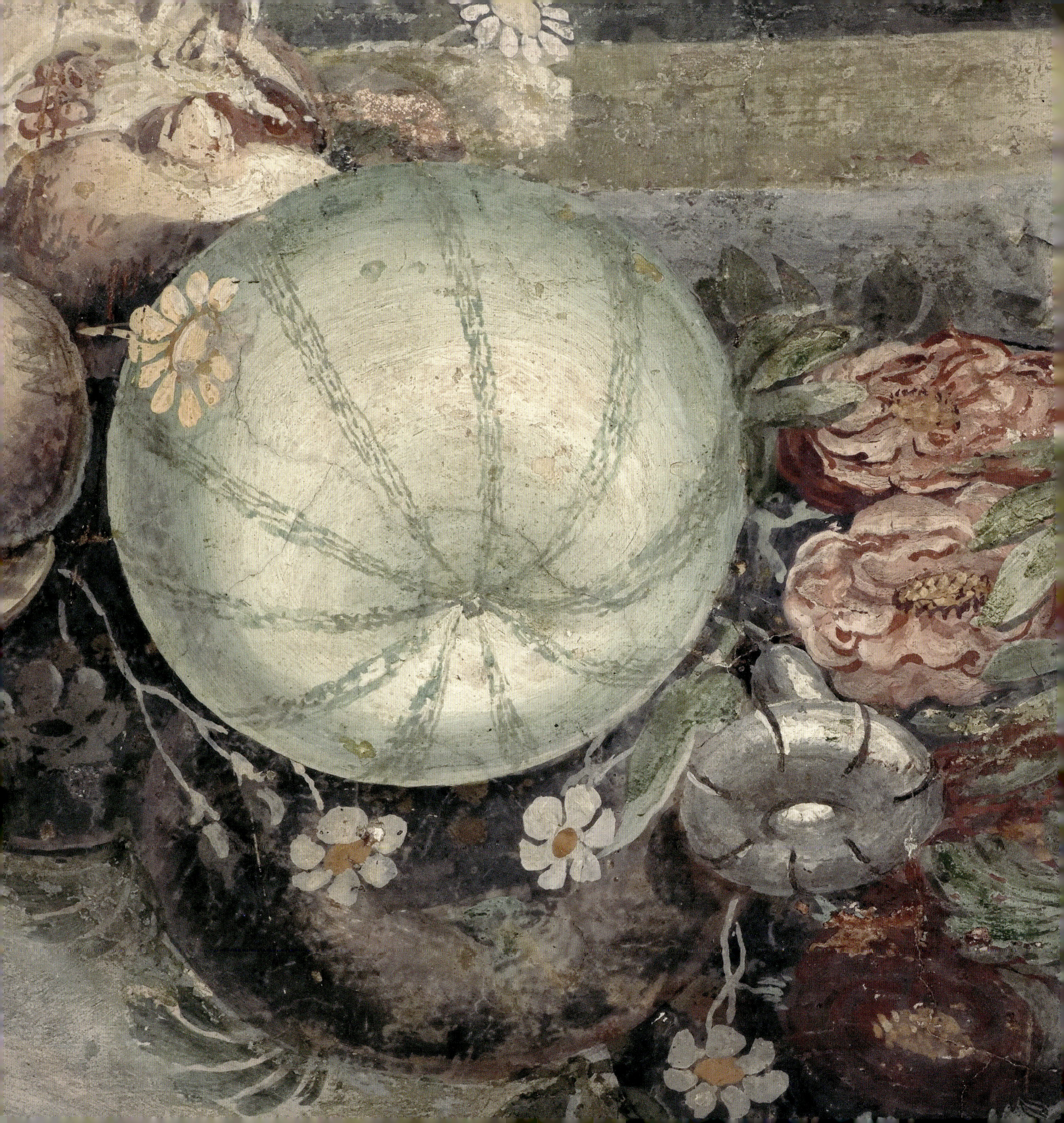

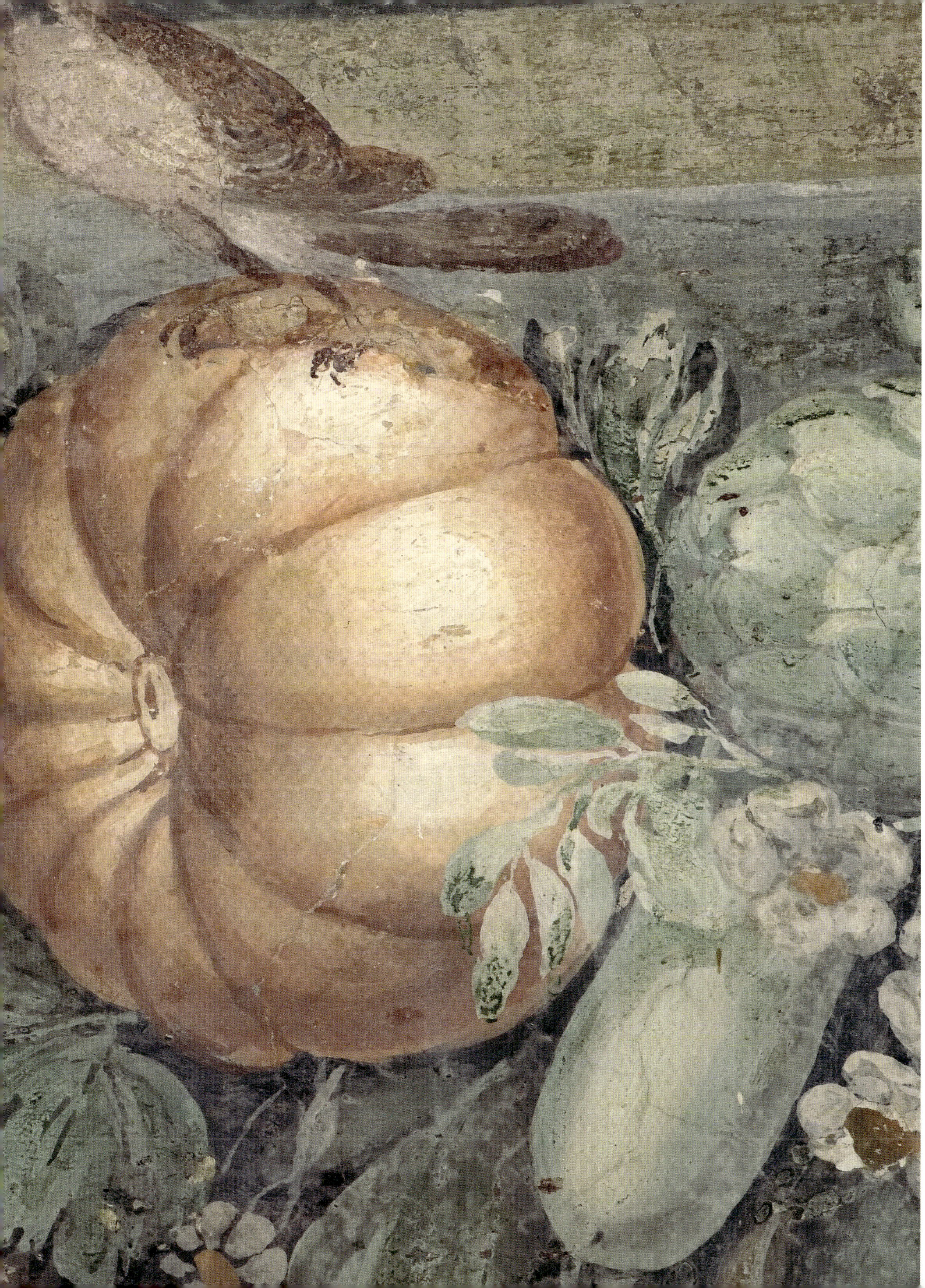

Detail with plant and floral festoons,
Hall of the Fountain, *piano nobile*

Pages 70–71
Workshop of Federico Zuccari,
Geometry and a *Liberal Art*, 1566–1567,
Hall of Nobility, *piano nobile*

THE GARDEN

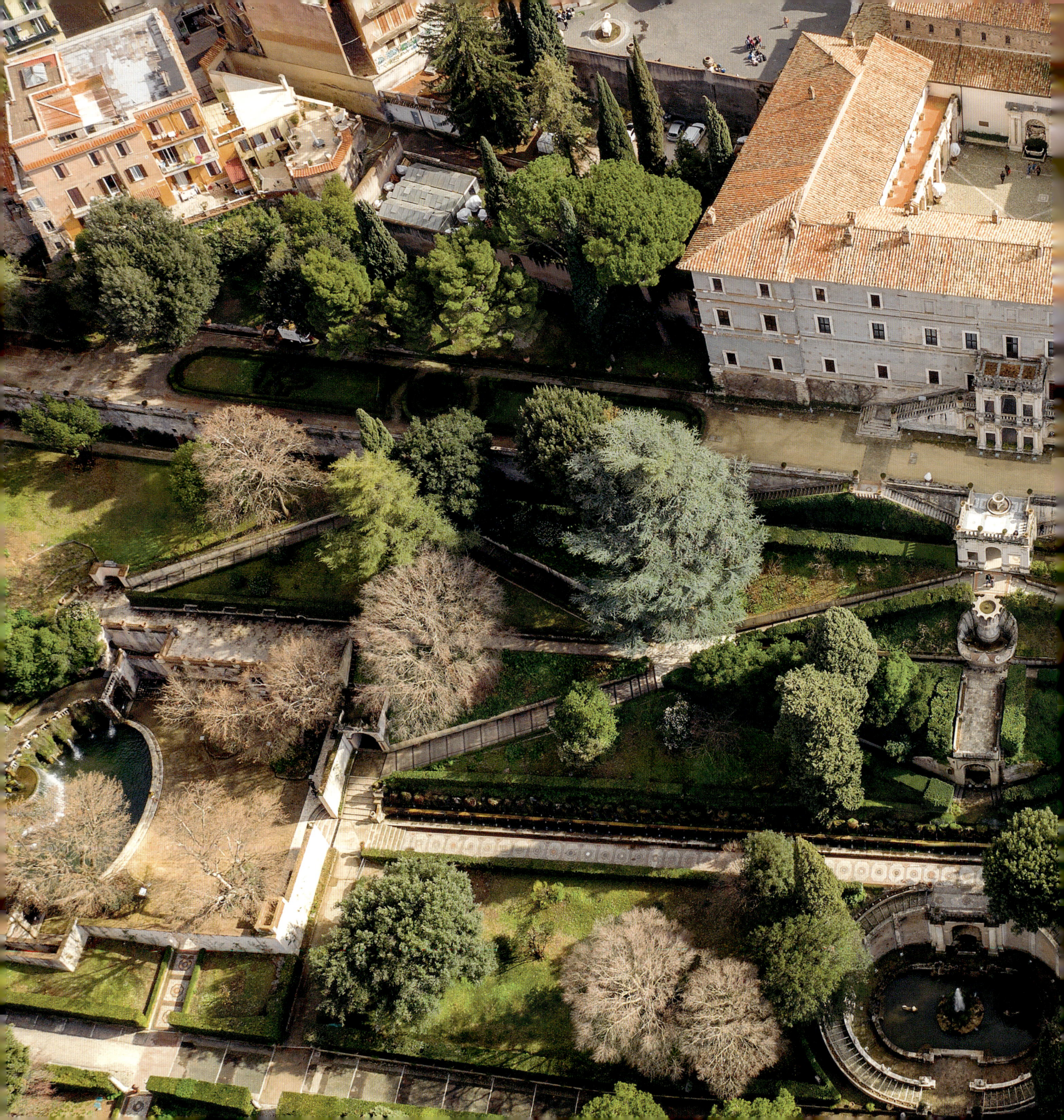

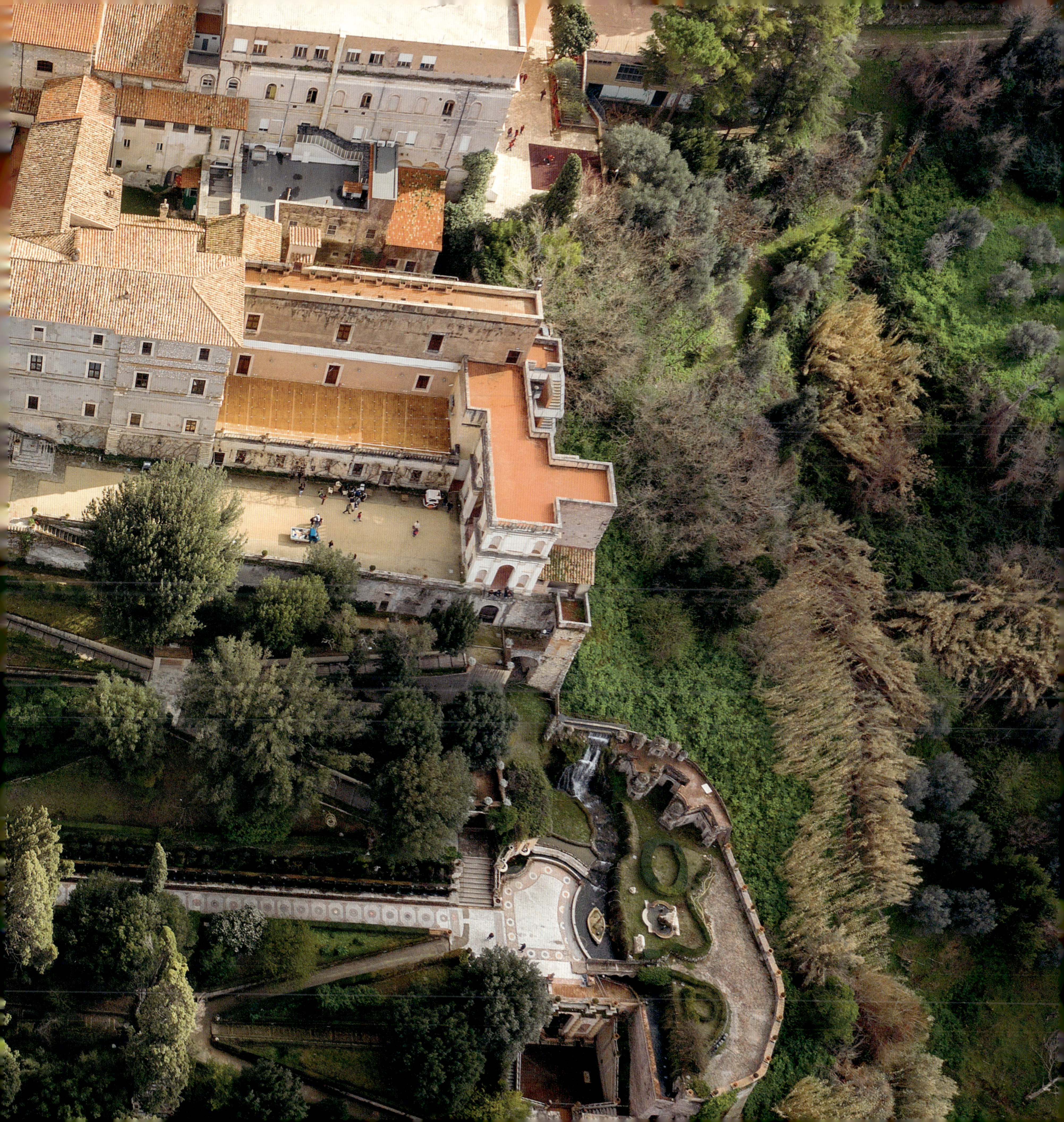

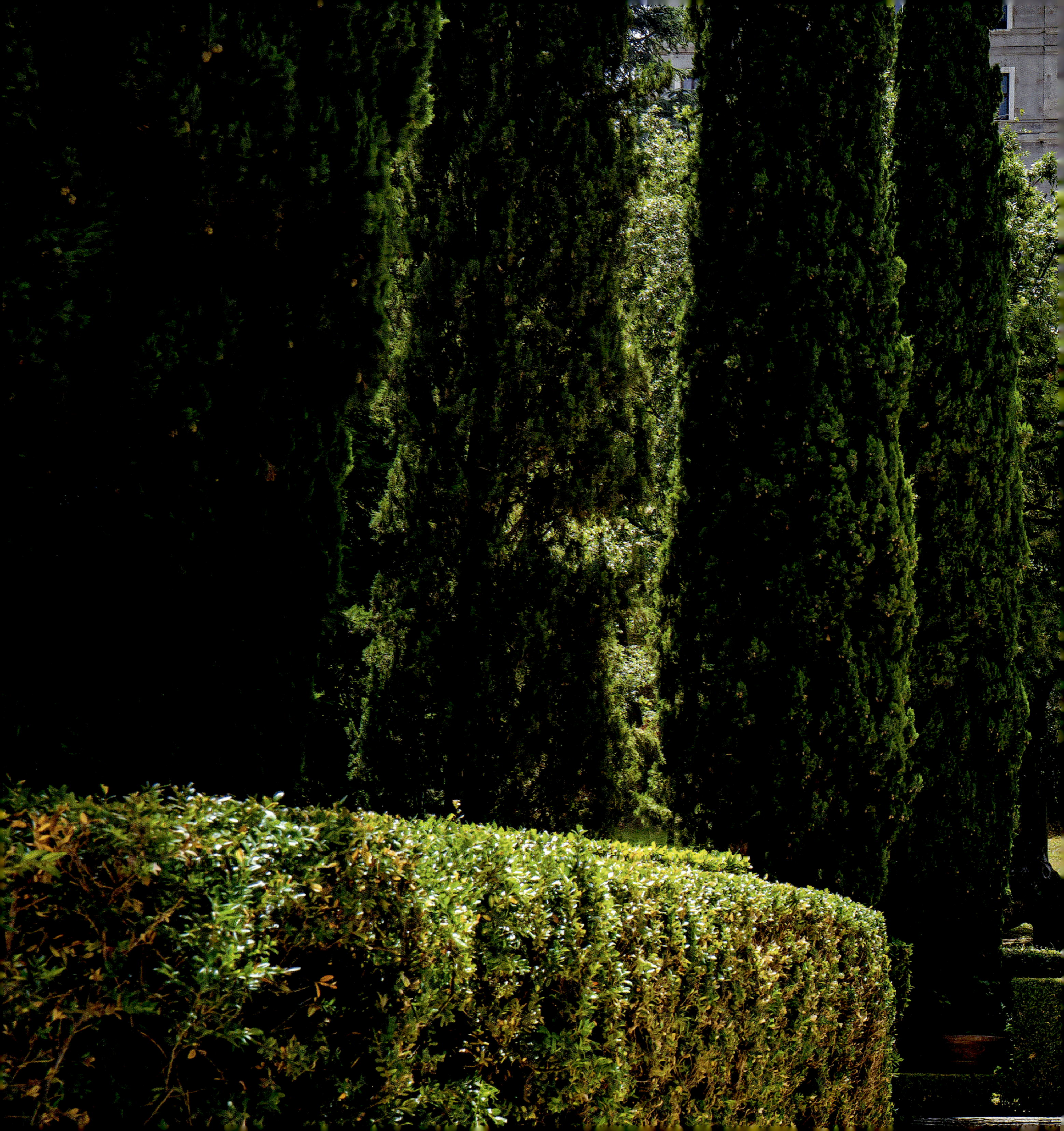

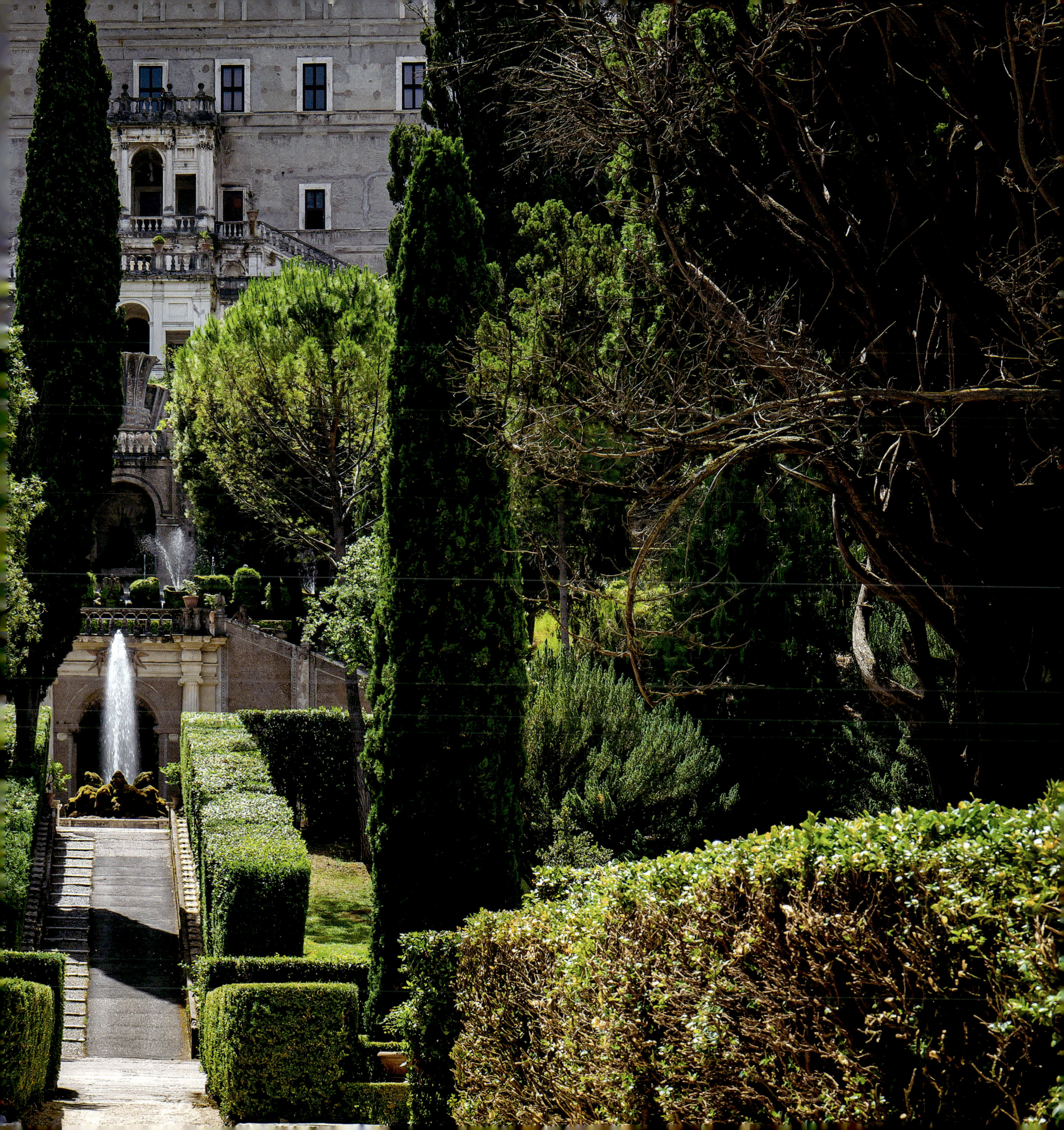

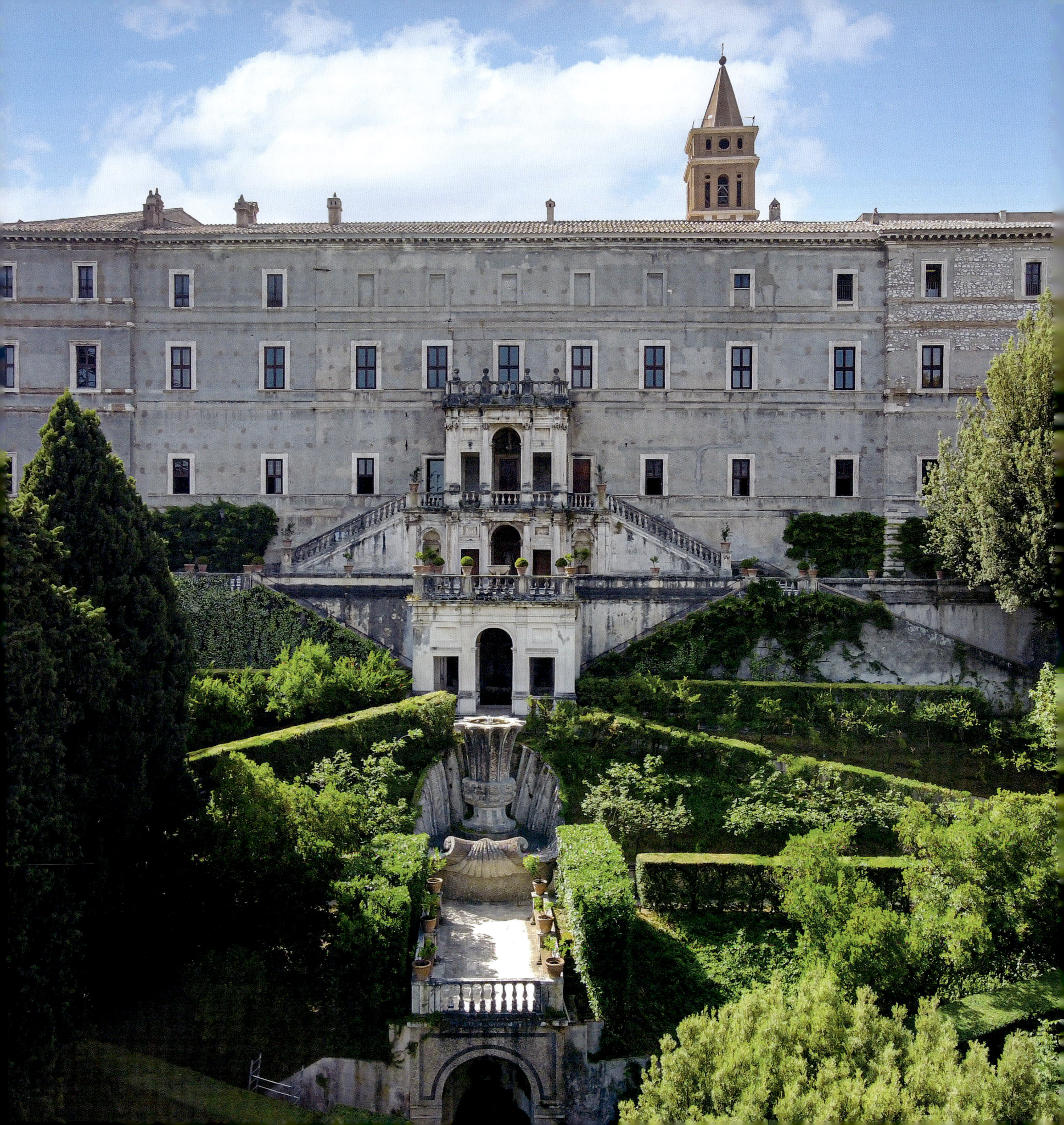

Garden

Parlan, fra le non tocche verzure, le cento fontane,
Parlan soavi e piane, come femminee bocche,
mentre sui lor fastigi, che il sol di porpora veste,
splendon, oh gloria d'Este, l'aquile e i fiordiligi.
Gabriele d'Annunzio

Not much of the present garden corresponds to its first depiction, as seen in Duperac's print from 1573, the year after the death of Cardinal Ippolito II. Nor does it correspond to the views by the same French artist later included in Lafrerinello's *Speculum Romanae Magnificentiae* (1575), and nor to the later seventeenth-century views, printed by Venturini, which mainly reflect the work carried out at the behest of Cardinal Alessandro d'Este. While the palace had already experienced centuries of decay and restoration, the garden, which was only partially completed during Ippolito's time and predominantly characterised by living nature, underwent continuous restoration and modifications.
Situated on a sloping terrain spanning approximately four hectares and featuring a height difference of about fifty meters, the garden is partly arranged along this slope and partly terraced. The eastern section is still supported by sixteenth-century substructures, which consist of structural elements repurposed from Roman times, comprised of brick walls connected by vaults.
In view of this characteristic and in compliance with the requests made by Ippolito II, the garden is accessible from the palace, located on the upper part, and from a gate on the lower side, allowing access for guests arriving from Rome. This entry provides them with their first grand and sublime view of the architectural and natural ensemble: the palace and the garden.
The complex water system, almost certainly attributable to Pirro Ligorio and derived from the architect's knowledge of the use of water in Hadrian's imperial villa, comprises an in-

View of the palace façade facing the garden
The three floors of the building, emphasised by stringcourses, rise on a basement that allows a view of the complex for those entering from the official entrance on Via del Colle. The windows are bordered on three sides by a linear travertine band. The double projecting loggia executed in the years 1566–1567, built in two orders (Doric and Corinthian), provides the upper apartment with an open space from which to admire the panorama and allows the entrance to the garden from the piano nobile, *via two scenographic flights of travertine side stairs. Underneath the loggia on the ground floor is a nymphaeum in which was the Fountain of Leda (the sculptural group is today preserved in the Galleria Borghese in Rome).*

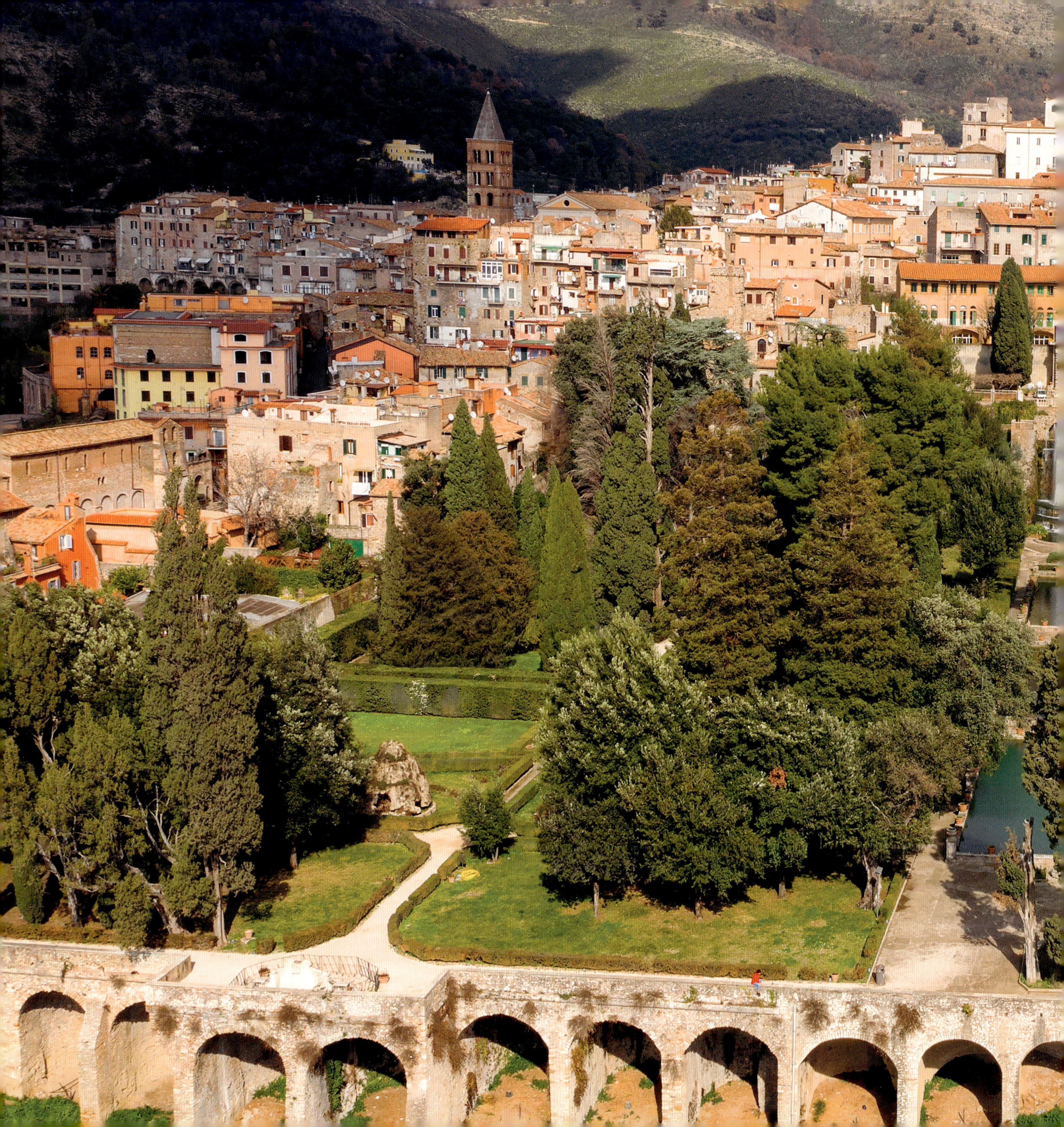

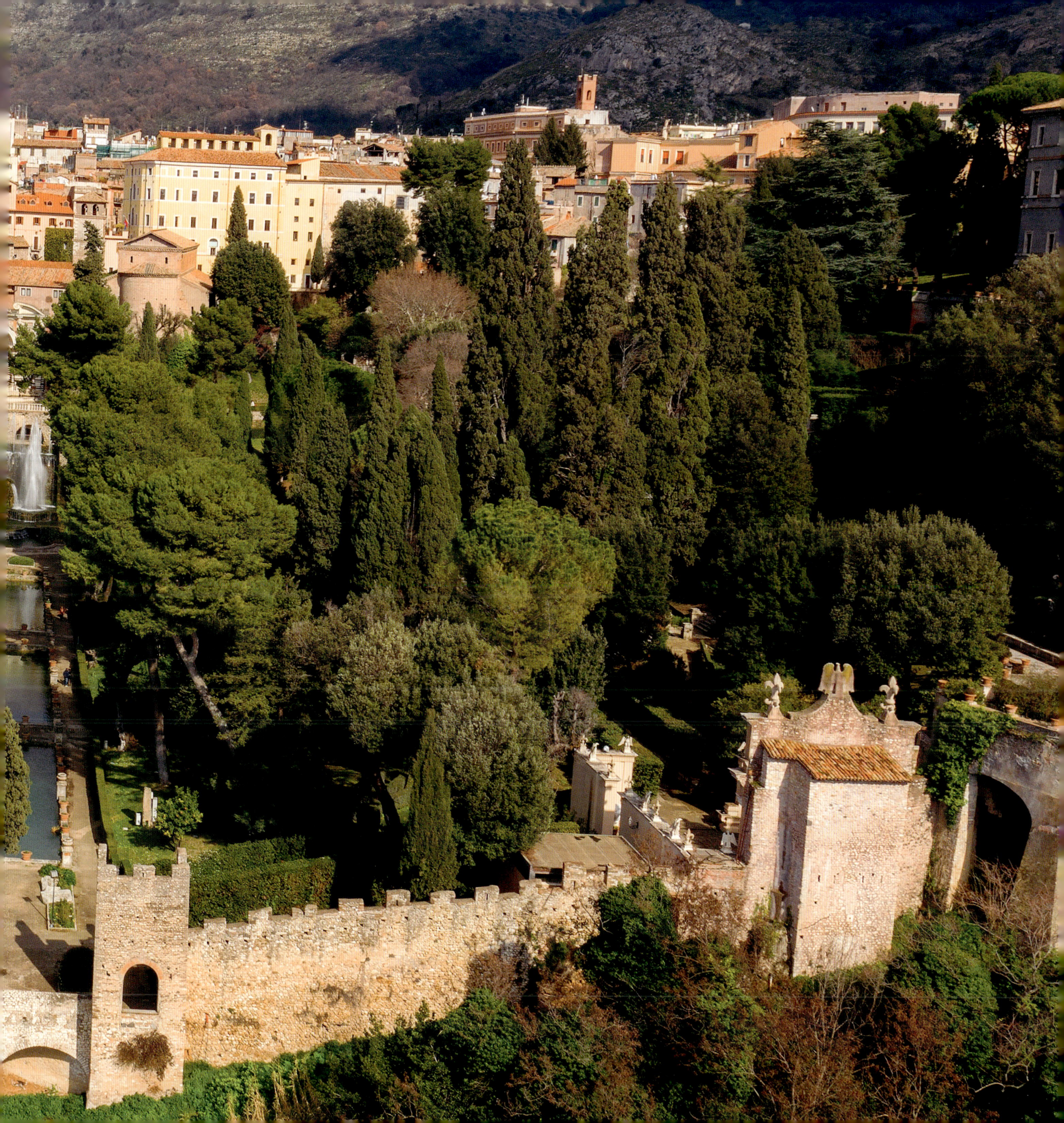

tricate underground network of conduits, tunnels and pipes fed, as it was originally, by the aqueduct from the Aniene river, measuring 282 metres in length. While this system was considered cutting-edge in the sixteenth century, the condition of the river water today causes rapid degradation of the fountains and the entire system. The existing filtering and purification treatment can only partially mitigate this issue, as it cannot prevent the significant accumulation of silt, which, when deposited, forms extensive and hardened concretions.

As per historical documents, the garden's original decoration was designed to include 51 fountains and nymphaea, featuring a total of 398 spouts, 364 jets, 64 cascades and waterfalls, 220 basins, tubs, and ponds, as well as 875 metres of water ducts and channels.

Grand Avenue

The first expansive, flat area, which extends beyond the Villa and was originally shaded by rows of elm trees, hosts the Fountain of the Tripod, situated in the section facing the residence itself. This fountain is a replica of the original Fountain of the Sea Horses, a composition that combines various elements such as a basin with marble supports (currently housed in the Louvre) and four horses supporting a small cup (now found in the Vatican Museums), all sourced from Hadrian's Villa.

Opposite the fountain stands the palace façade, which remains unfinished but features an elegant loggia flanked by staircases. This loggia incorporates two superimposed orders, the Doric and Corinthian. The upper level is accessible from the Hall of the Fountains and Hall of Hercules, while the lower-level forms a portico housing the Fountain of Leda, named after the sculptural group of *Jupiter transformed into a Swan* that originally adorned it. This group, which is now in the Galleria Borghese in Rome, has been replaced by a headless Minerva.

Continuing eastward along the promenade, one passes an elevated area where the Cardinal's Secret Garden once stood, eventually reaching the Fountain of Europa, erected in

Fountain of the Tripod
This is a copy of the one now in the Louvre, placed here by Attilio Rossi around the 1930s.

Pages 82–83
Aerial view of the garden and the historical centre of Tivoli

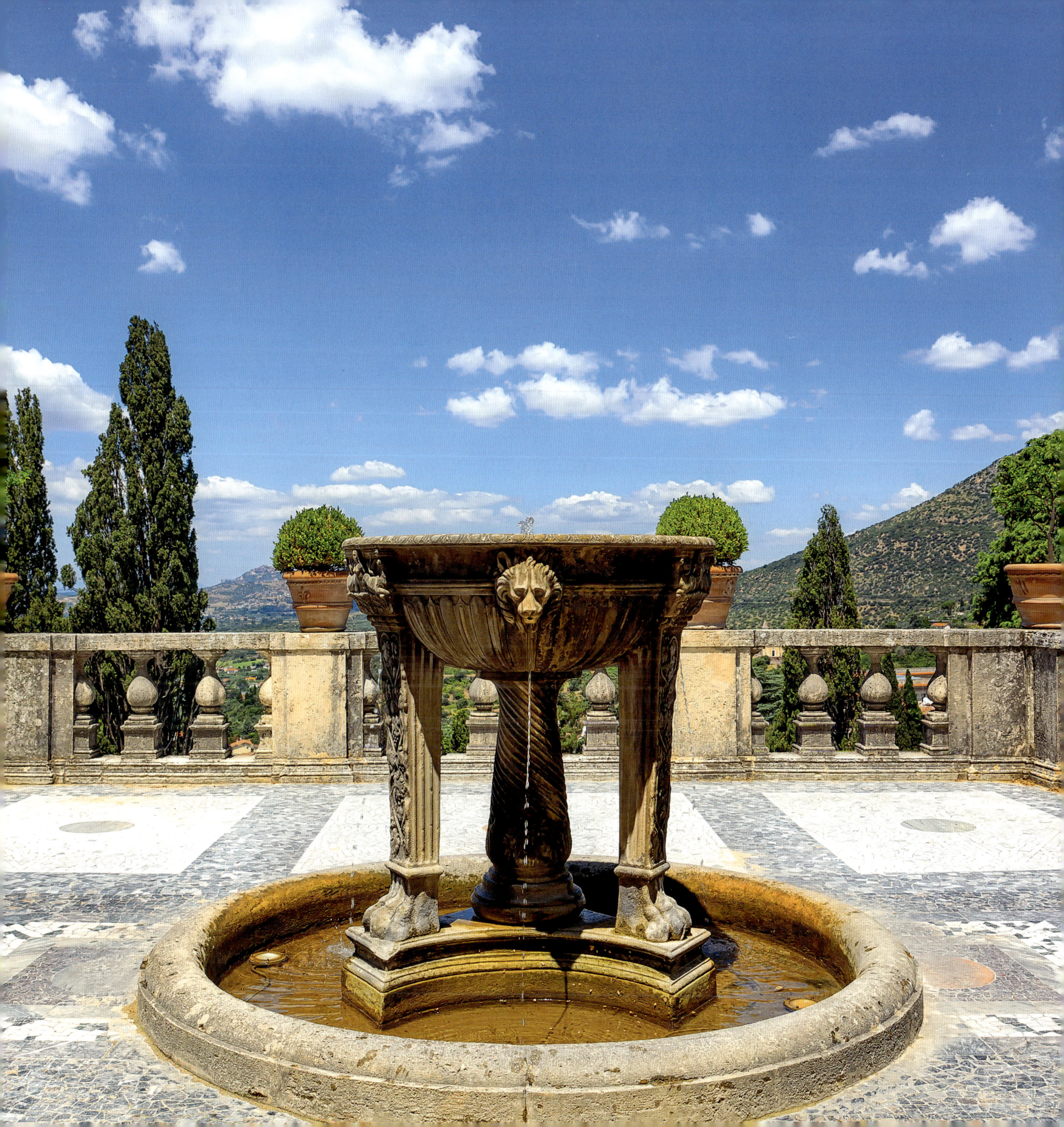

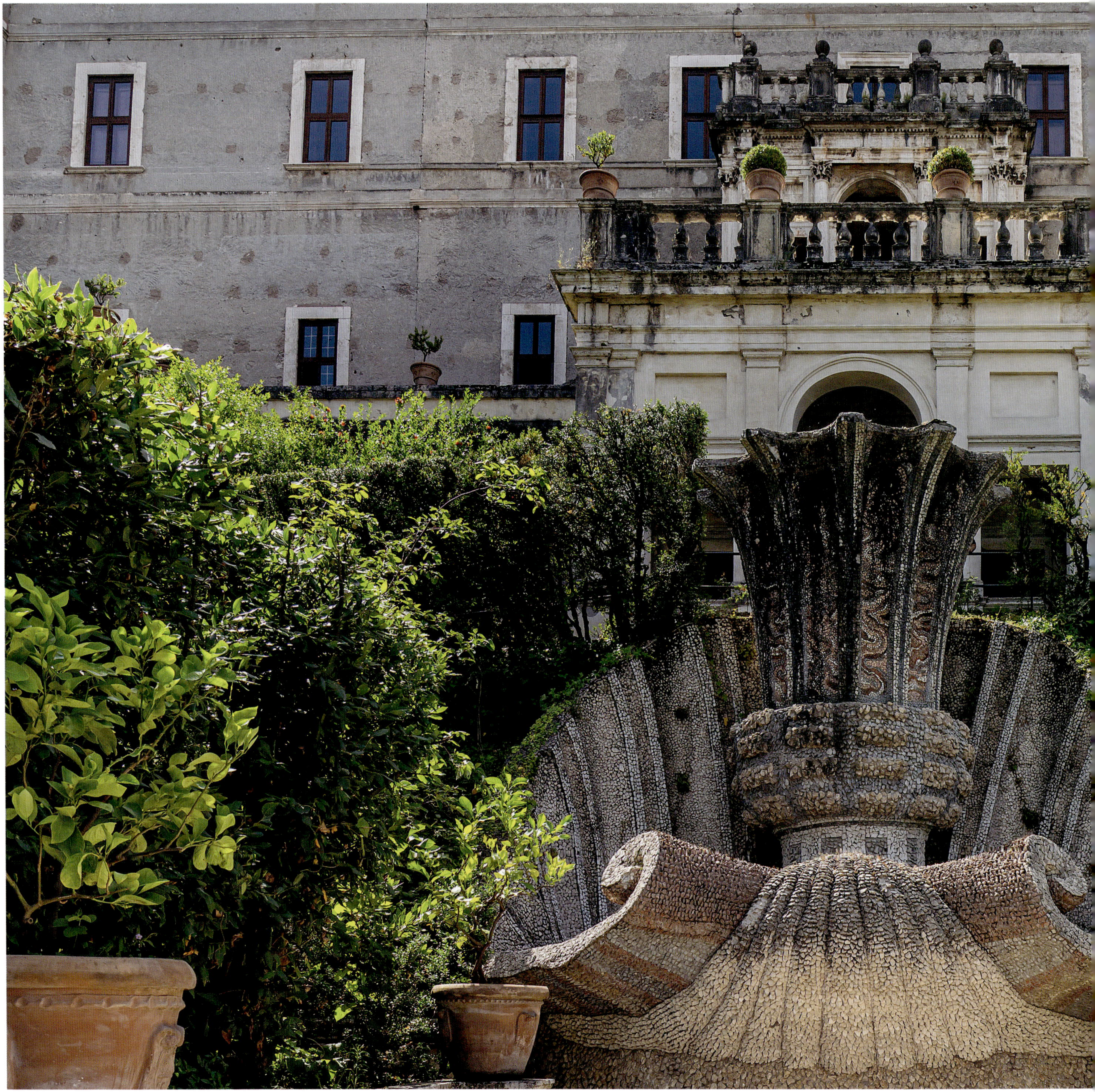

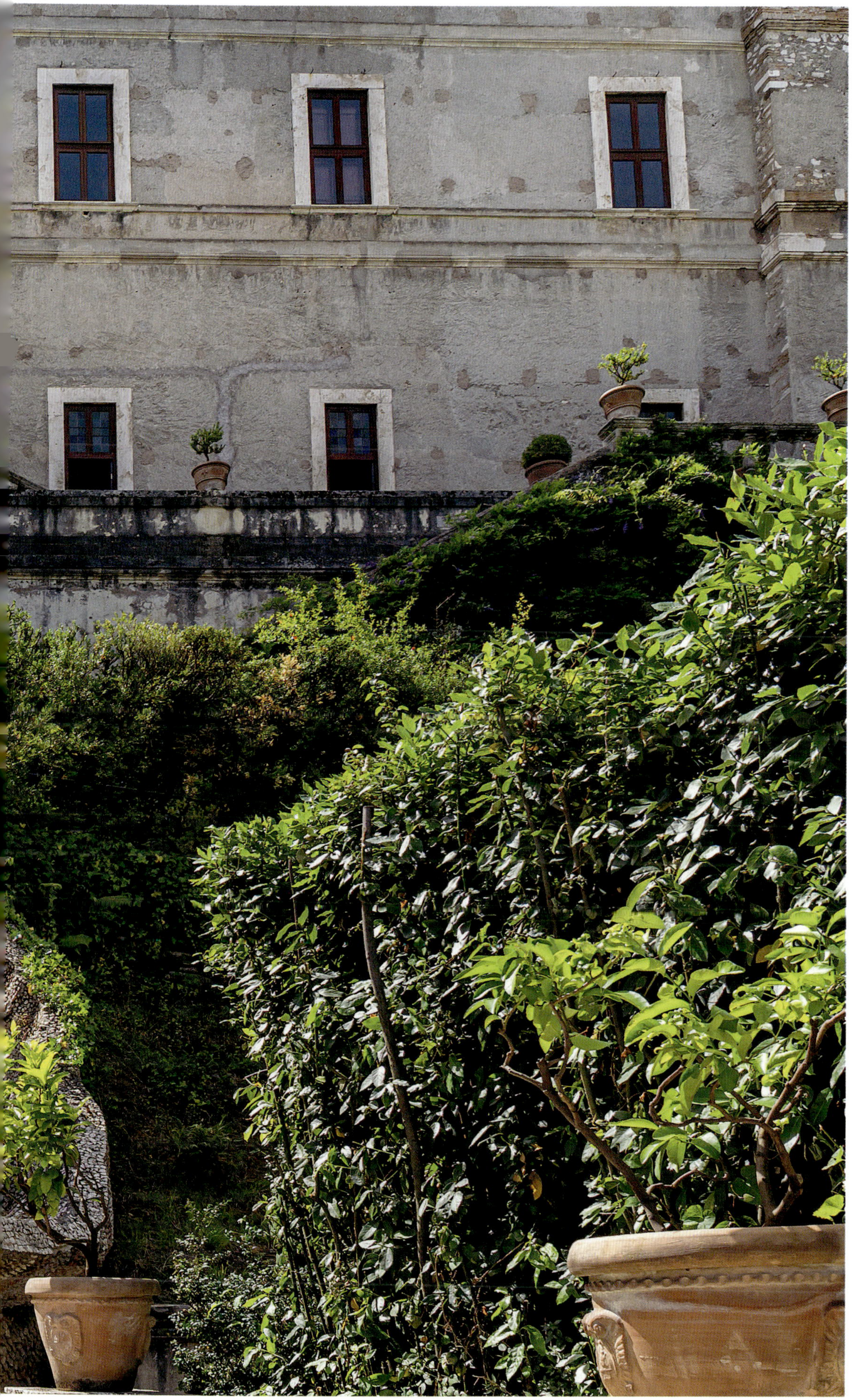

Gian Lorenzo Bernini,
Bicchierone Fountain, 1661
The Loggia of Pandora can be seen in the background.

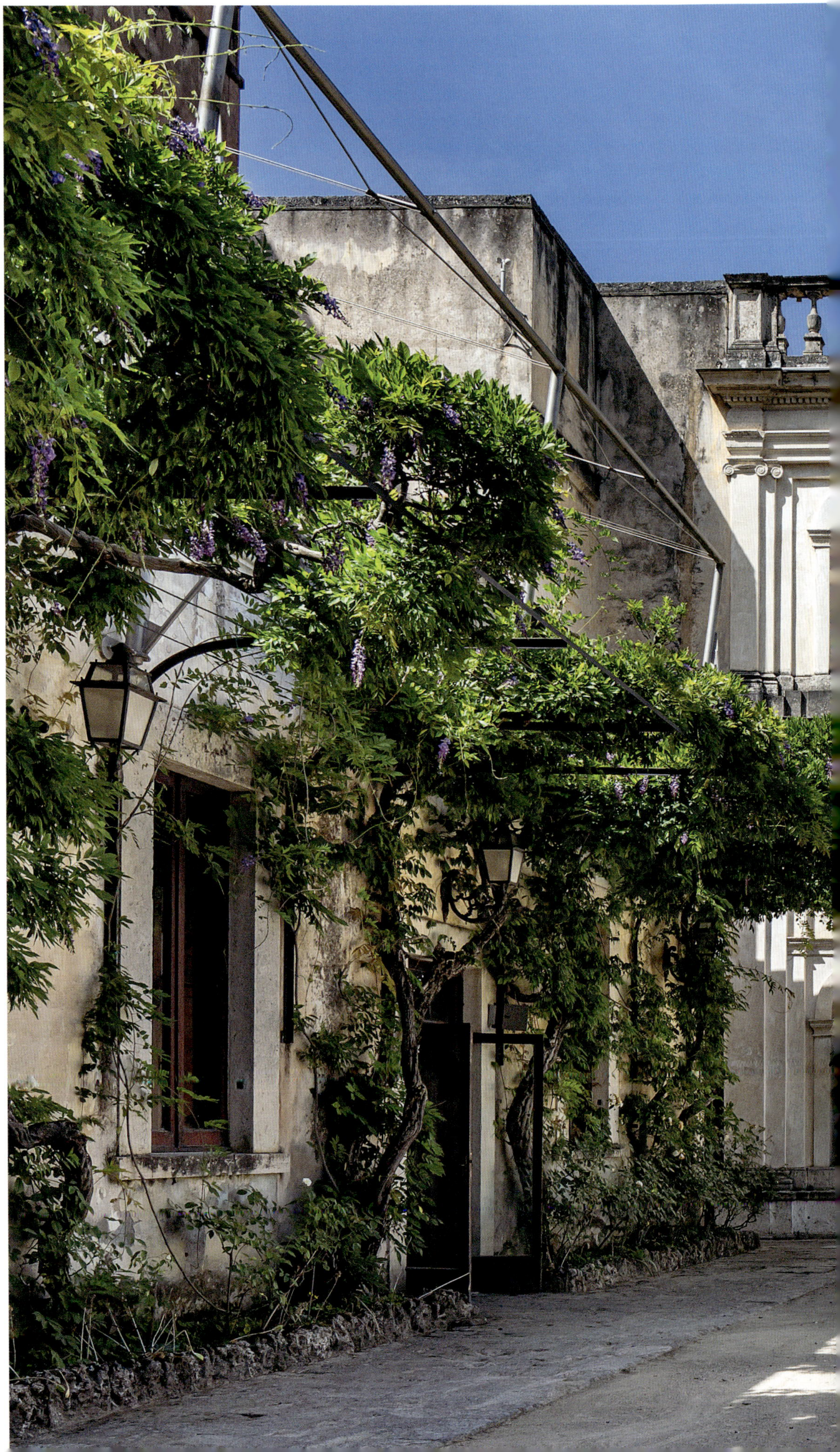

Pirro Ligorio, Gran Loggia
or Cenacolo, 1568–1569

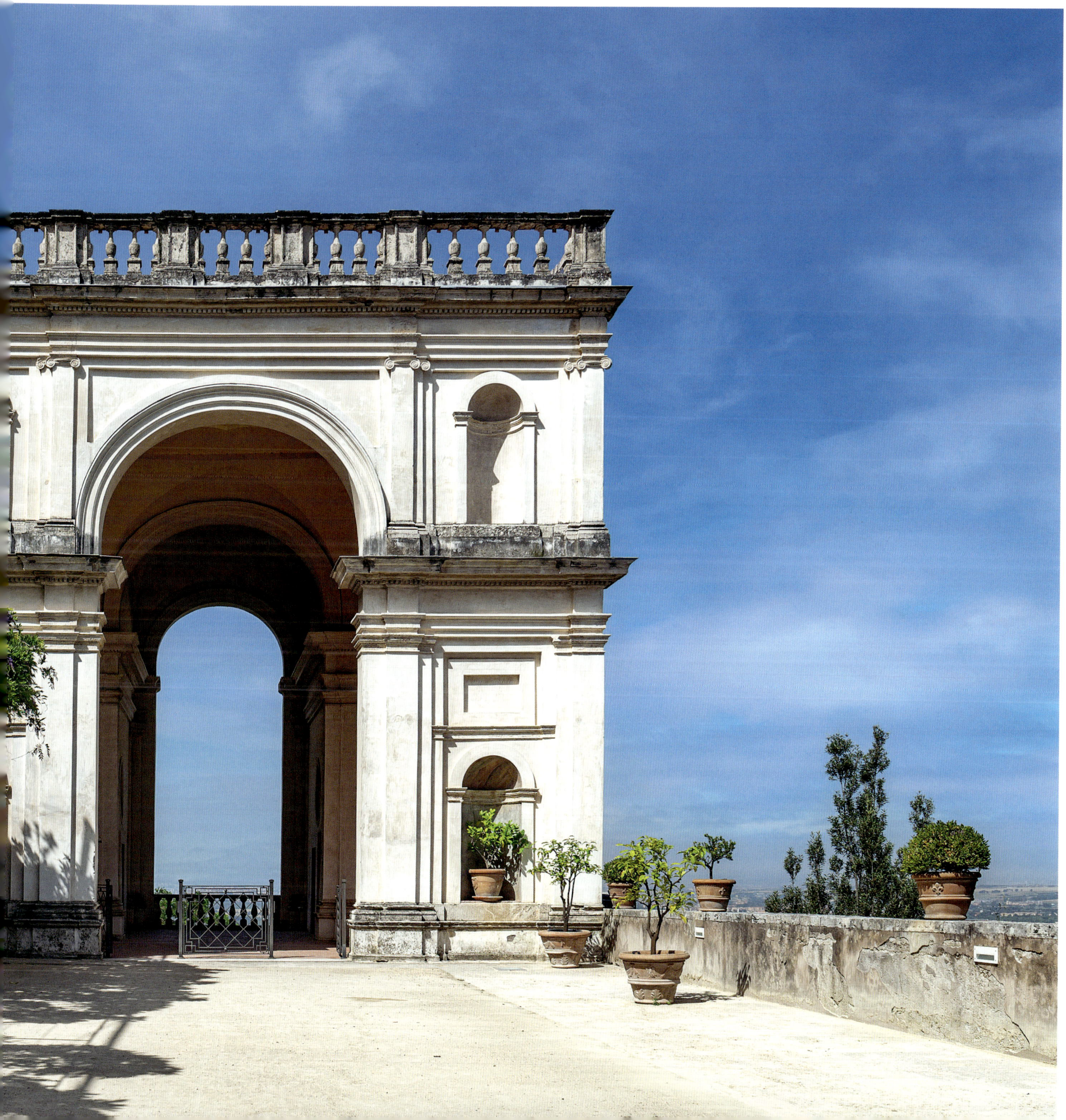

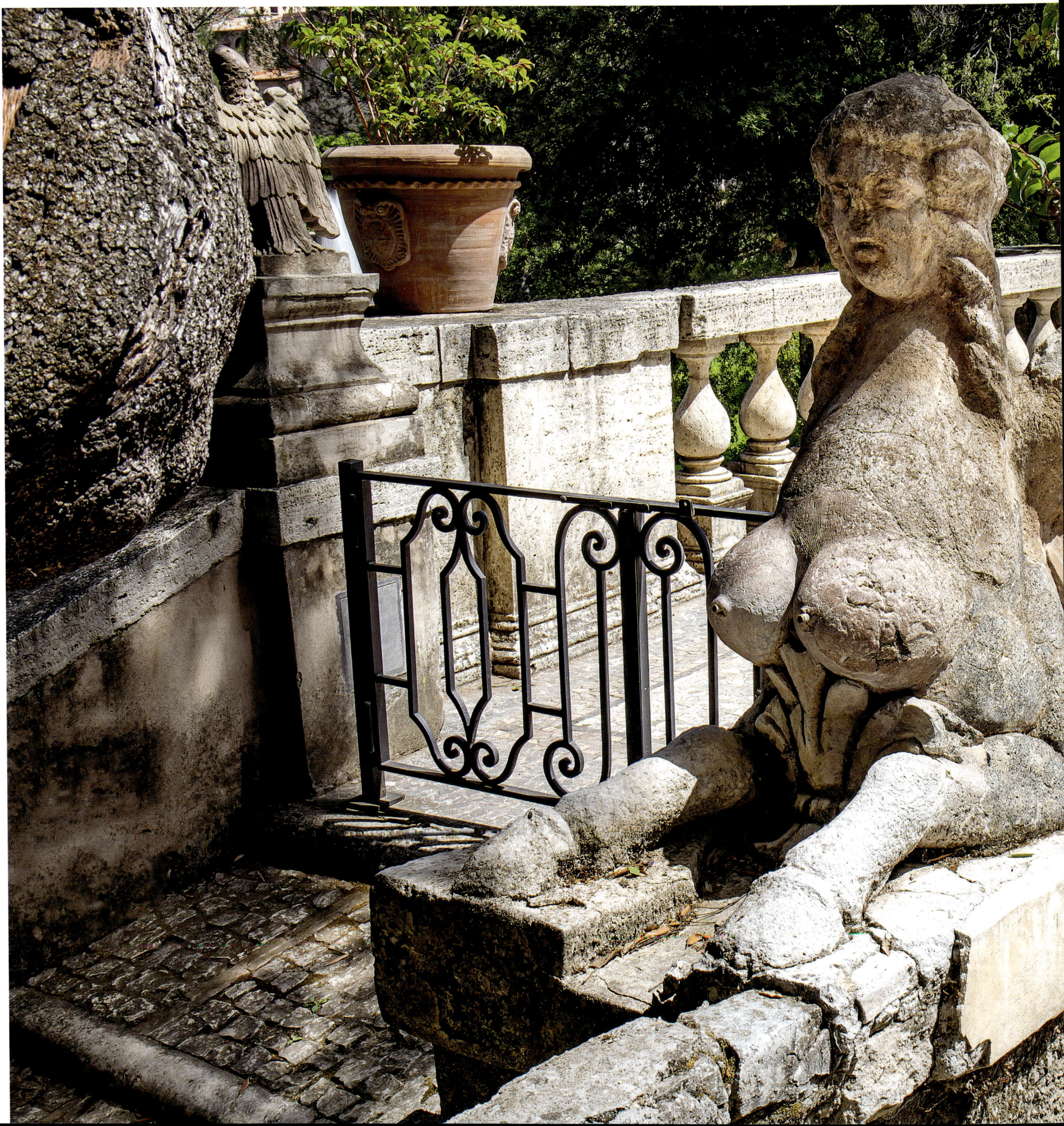

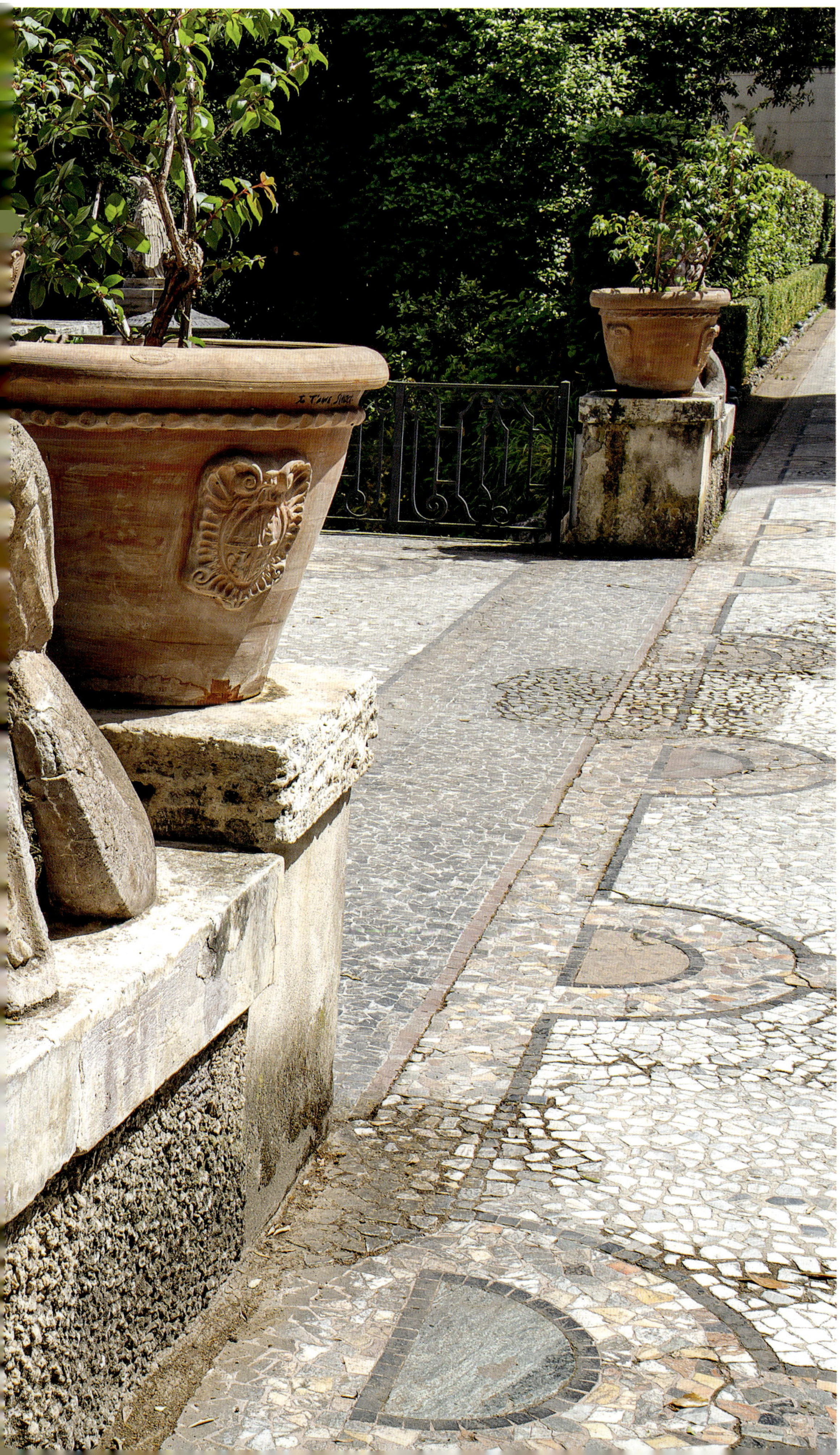

Sphinx in the Avenue of the Hundred Fountains

1671. The fountain takes its name from a sculptural group from antiquity, *Europa embracing the bull*, which used to adorn it and which, like many others, is now in Rome, in Villa Albani. In addition to this group, a large marble basin in which water was collected has also been lost. On the opposite side, after passing through the majestic wisteria pergola below the Terrace of the Pallacorda, one can admire the splendid panorama stretching as far as Rome, framed by the large archway of the Gran Loggia or Cenacolo, built to a design by Pirro Ligorio between 1568 and 1569.

Below the Grand Avenue is the Passeggiata del cardinale, or Cardinal's Walk, a path flanked on one side by the supporting structures of the Avenue's terracing. These structures, mostly concealed by brickwork, were embellished with several nymphaea, among which the central one, the Loggia of Pandora, presents a façade designed by Serlio and, within, a niche adorned with a Pandora flanked by two statues (now in the Capitoline Museums), all from Hadrian's Villa.

At the ends of the Promenade, other nymphaea are embedded into structural niches. To the west, you'll find the Grottoes of Aesculapius and Hygieia, both adorned with mosaic decorations and once housing statues of Aesculapius (now in the Louvre) and Hygieia (now in the Vatican Museums). To the east is the final nymphaeum, the Grotto of Diana, with a cruciform design that is only visible from the outside due to its fragility. The vaults and walls of this grotto were decorated between 1570 and 1572 by Paolo Calandrino and clearly draw inspiration from antiquity, particularly from Hadrian's Villa. They feature stucco, stone, enamel, glass paste, and shells, sometimes interspersed with majolica ceramics, which frame large bas-relief panels depicting mythological scenes on the side walls. The original sculptural decoration in the rear niche, consisting of a Diana, two statues of Amazons, and a Minerva from Hadrian's Villa, is now housed in the Capitoline Museums. The floor decoration in majolica, executed by craftsman Bernardino de' Gentili from Aversa, is only partially preserved.

One can reach the lower level from each of the paths branching off from the Cardinal's Walk, marked by the presence of the Bicchierone Fountain. This fountain's design is based

Fountain of Bacchus, 1568–1569

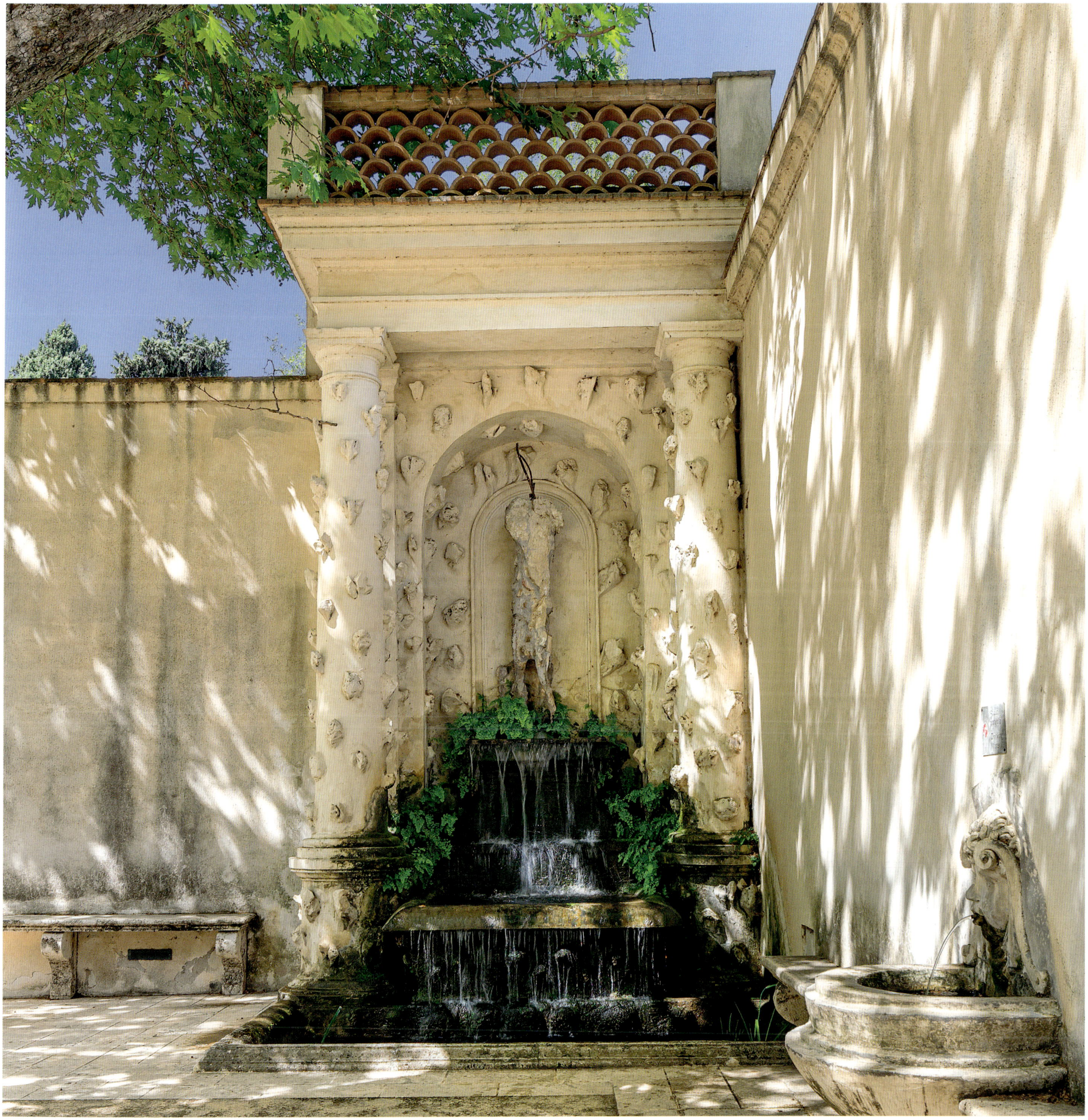

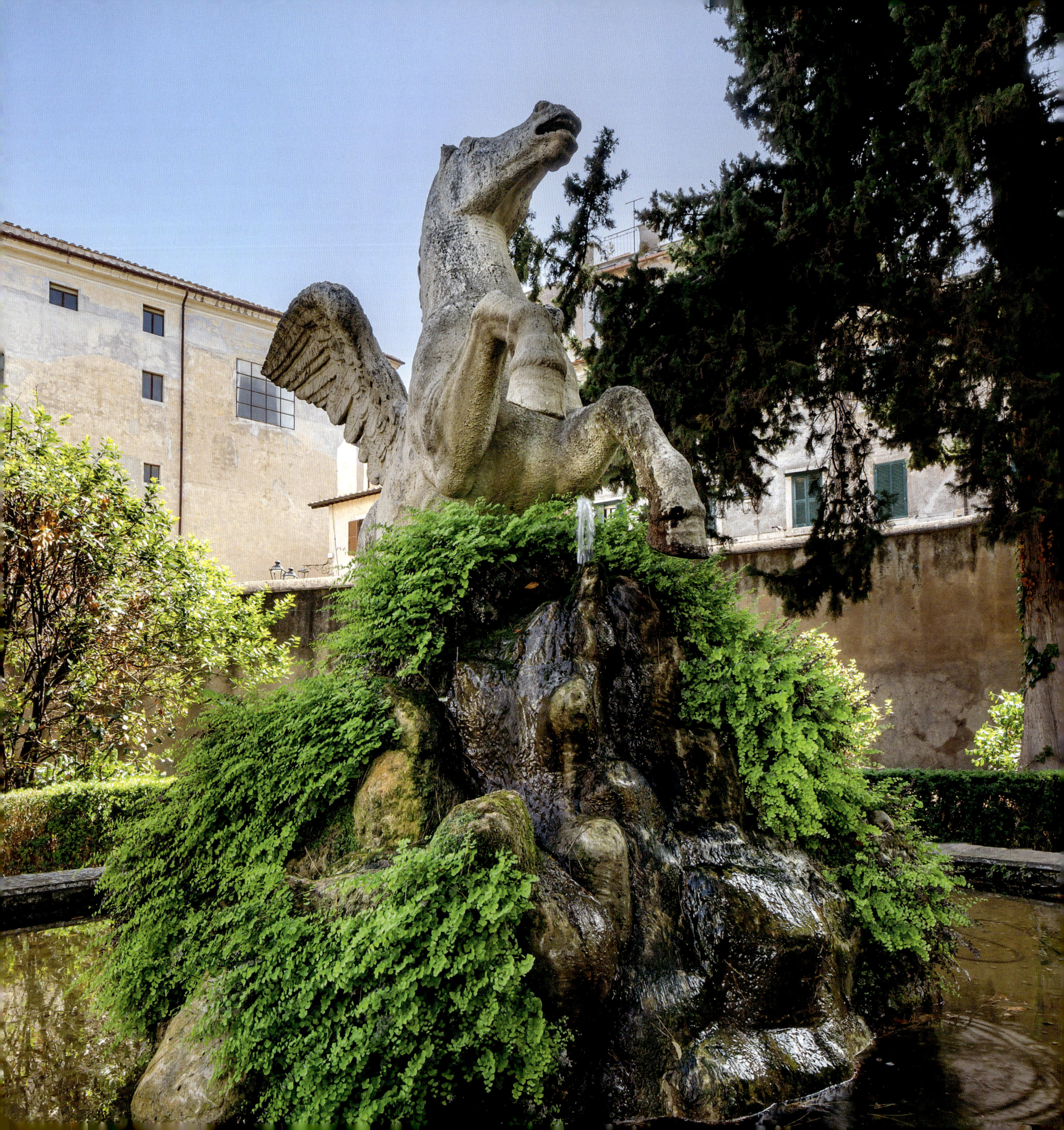

on a sketch by Gian Lorenzo Bernini, as described in an autograph letter he wrote to Cardinal Rinaldo d'Este on 15 June 1661: "I was taken to Tivoli, arrived at eleven o'clock, and because I found some things that had not been done to my taste, I had some boards made with stones and mortar of that which was in the bottom of the fishponds and worked until I was satisfied, and everything was adjusted so that I could test the water, which both in quantity and in its play was more to my taste than I had imagined, and the magnificence of the site is so great, and makes such a stir that I believe it will satisfy your learned taste. Before leaving, I ordered M. Filippo to make everything of good material that, as I said, was made of boards, stones and mortar, which work should be done in six days". The fountain, originally more correctly called the Fountain of the Lily, comprises two indented chalices (the lilies of the family coat of arms) stacked atop each other, emerging from the valve of a scallop shell. The entire surface is finely decorated with polychrome stone tesserae and from the inside of the upper chalice emerges a mighty gush of water whose flow rate was reduced by Bernini himself in order not to obstruct the perspective view of the areas behind. Immediately below this level lies one of the most enchanting and spectacular areas in the entire garden. Here, you can admire the Fountain of Ovato (or Tivoli) and the Fountain of Pegasus in the eastern part. In the central section, marked by the majestic Avenue of the Hundred Fountains, there are the Grotto of Venus, the two Fountains of Bacchus, and the Grotto of Pomona. Moving further along the central part, you'll encounter the Fountain of Hercules, while in the western sector, there are the Grotto of Flora, the Rometta Fountain, the Proserpine Fountain, and the Owl Fountain.

Fountain of Pegasus

Closely associated with the Fountain of Tivoli, also known as the Fountain of Ovato, it is an integral part of Ligorio's celebratory program. Tivoli, with the arrival of Ippolito, became a hub for the arts. In this context, the winged horse symbolizes the mythological tradition of

Fountain of Pegasus,
second half of the sixteenth century

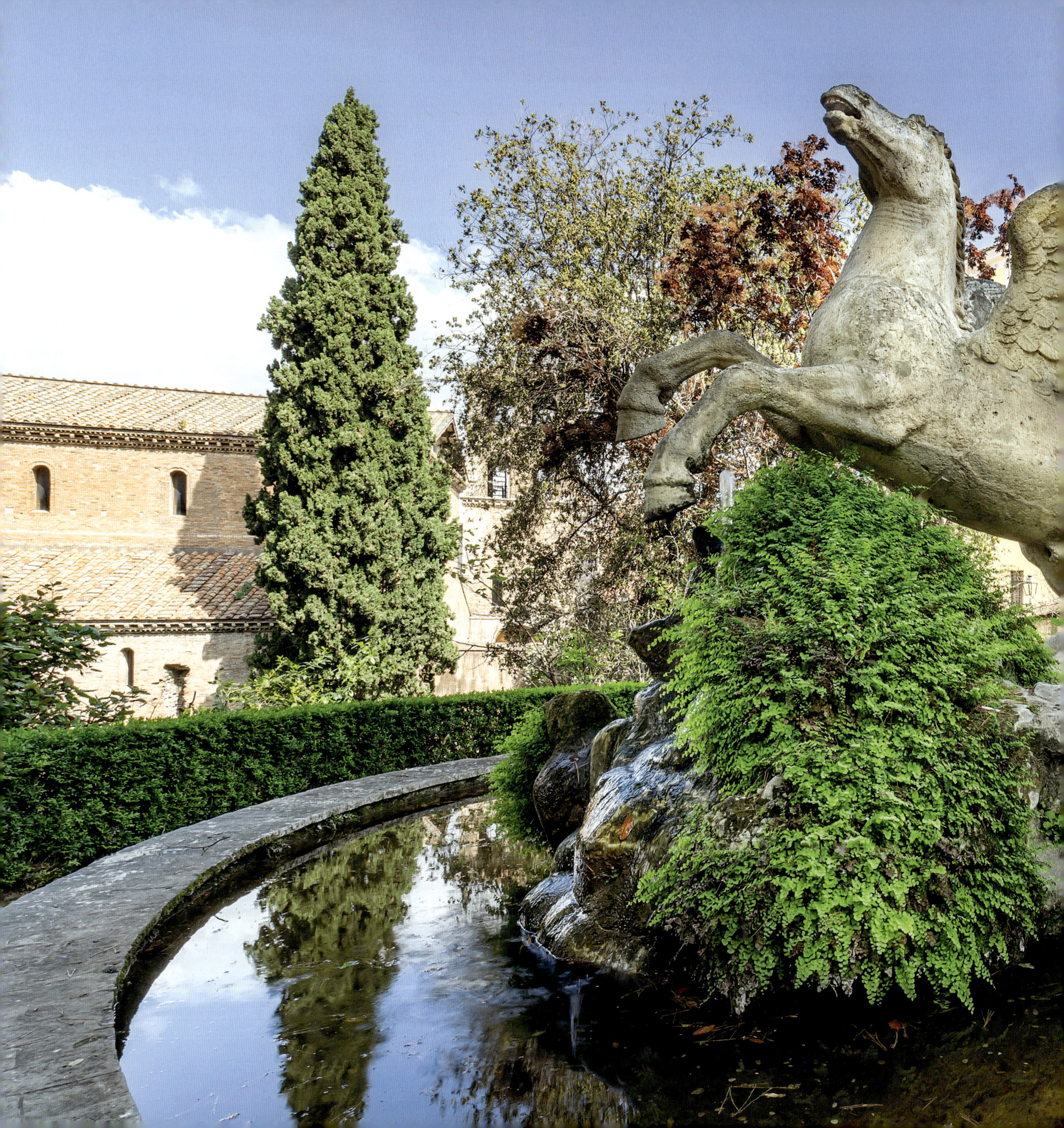

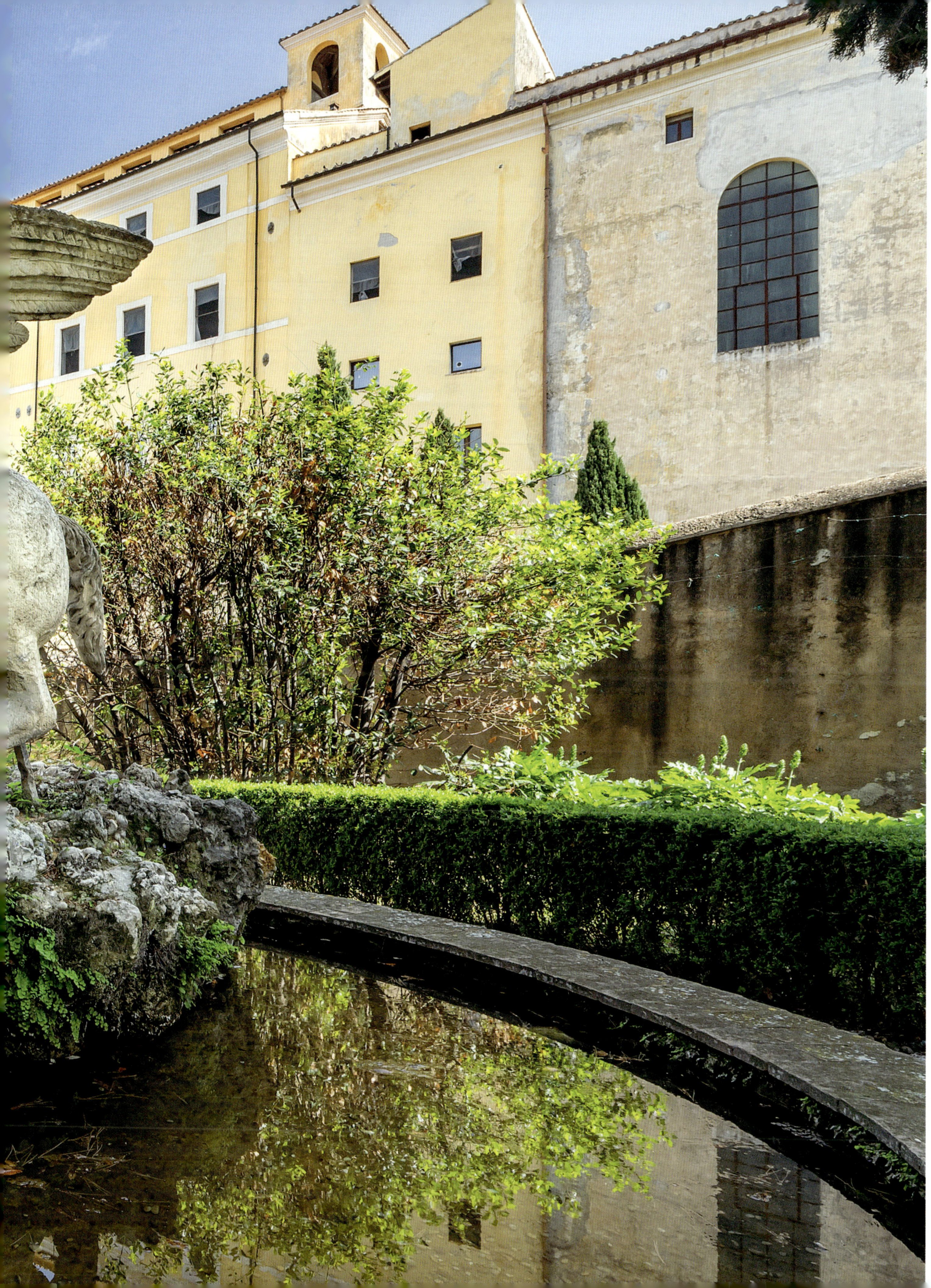

Fountain of Pegasus, second half
of the sixteenth century

Pegasus flying towards the Mount of the Muses. This sculpture likely postdates the contract signed in November 1568 by Cardinal Provost of Tivoli, Visdomino Visdomini, with two sculptors in Rome: Gillis van den Vliete (Gilio Vellita), a Frenchman, and Pierre de la Motte (Pietro della Motta), a Fleming. Both were *huomini periti dell'arte* – experts in the art – and were commissioned to make three travertine sculptures in conformity with the design, *che è stato datto da m. Pirro Ligurio* – which was given by messer Pirro Ligurio – representing the Tiber, the She-wolf with Romulus and Remus and an Eagle.

Fountain of Tivoli or of Ovato

The lower-level area hosts one of the most stunning fountains, an early example of 'water theatres' designed to stage aquatic performances in various forms. This is the Fountain of Tivoli or Ovato, attributed to Ligorio's design and constructed in 1565, as evidenced by an illustration in the first Sala Tiburtina depicting its construction phase. It is fully described in a verse written after 1569 by Oberto Foglietta, a historian serving Cardinal Ippolito II, and later Cardinal Luigi. The fountain holds a central position in the garden's layout, serving as the initial point for the aqueduct that supplies water from the Aniene and feeds all the other fountains in the garden.

Already described by sources as "the principal of all the fountains in this garden, & perhaps of all Italy", the fountain has been the object of admiration over time. Its central position is emphasizsd by its location at the start of the first of the two grand "waterways" connecting the garden's most significant features. The architectural structure can be admired from the inside of an enclosure that was intended, as Ligorio's plan envisioned, for "outdoor beds in summer on which to rest during the day" while enjoying the cooling effect of the water jets. The fountain's design features multiple levels, starting with a lower exedra nymphaeum, referred to as the 'theatre of water', surrounded by grottoes and artificial rock formations. The exedra, with alternating arches and niches, houses ten statues of Nereids designed

Pirro Ligorio, Fountain of Ovato,
or of Tivoli, 1565–1569
Work began in 1565 on behalf of Tommaso da Como with the subsequent contribution of the fountain-maker Curzio Maccarone. Raffaello da Sangallo was responsible for the travertine elements and Paolo Calandrino for the mosaics and stuccoes. The spectacular nymphaeum is one of the first "water theatres". Behind the building lies the conduit that carries water from the Aniene River, which from here divides into the different distribution lines of the garden.

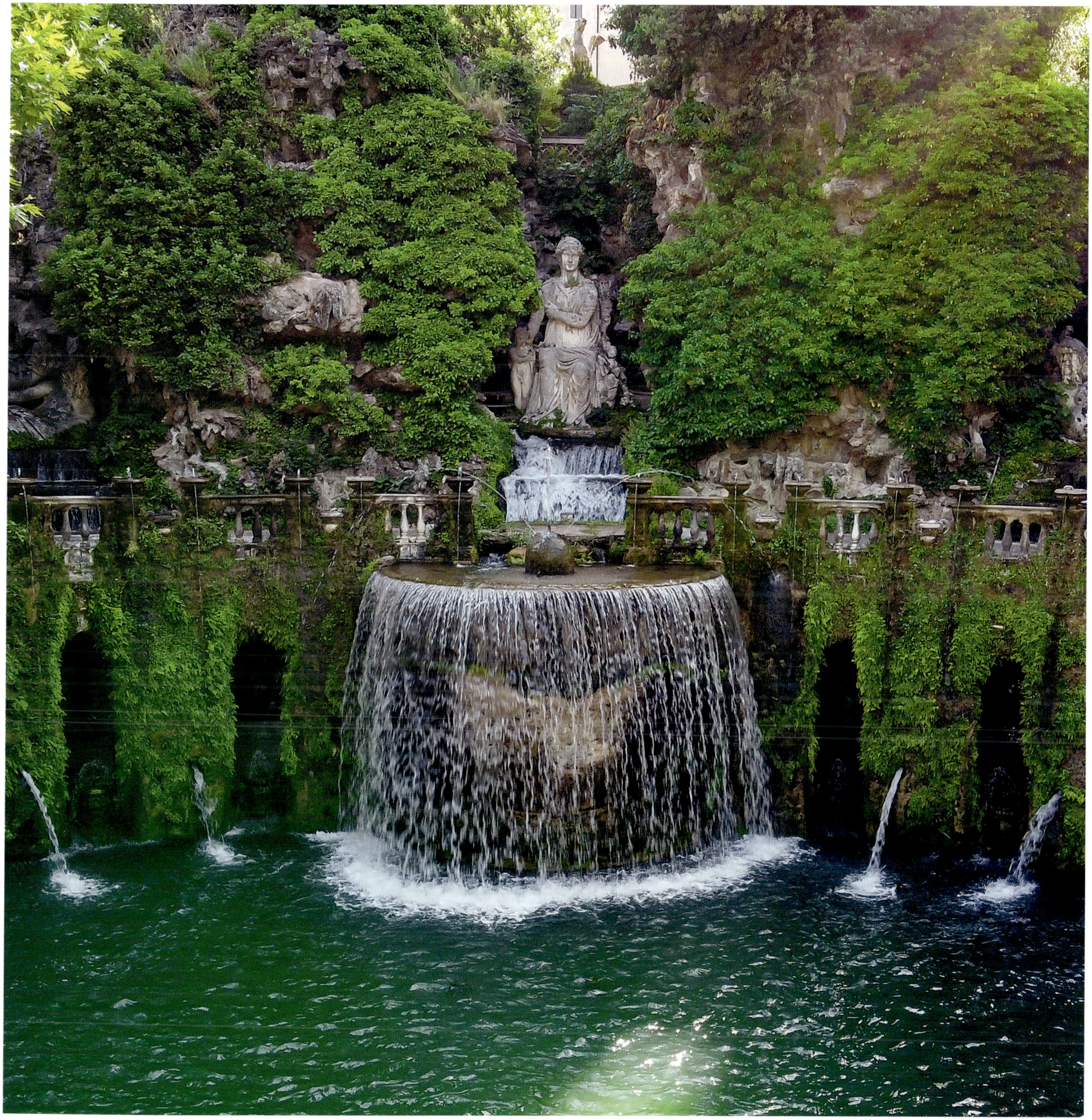

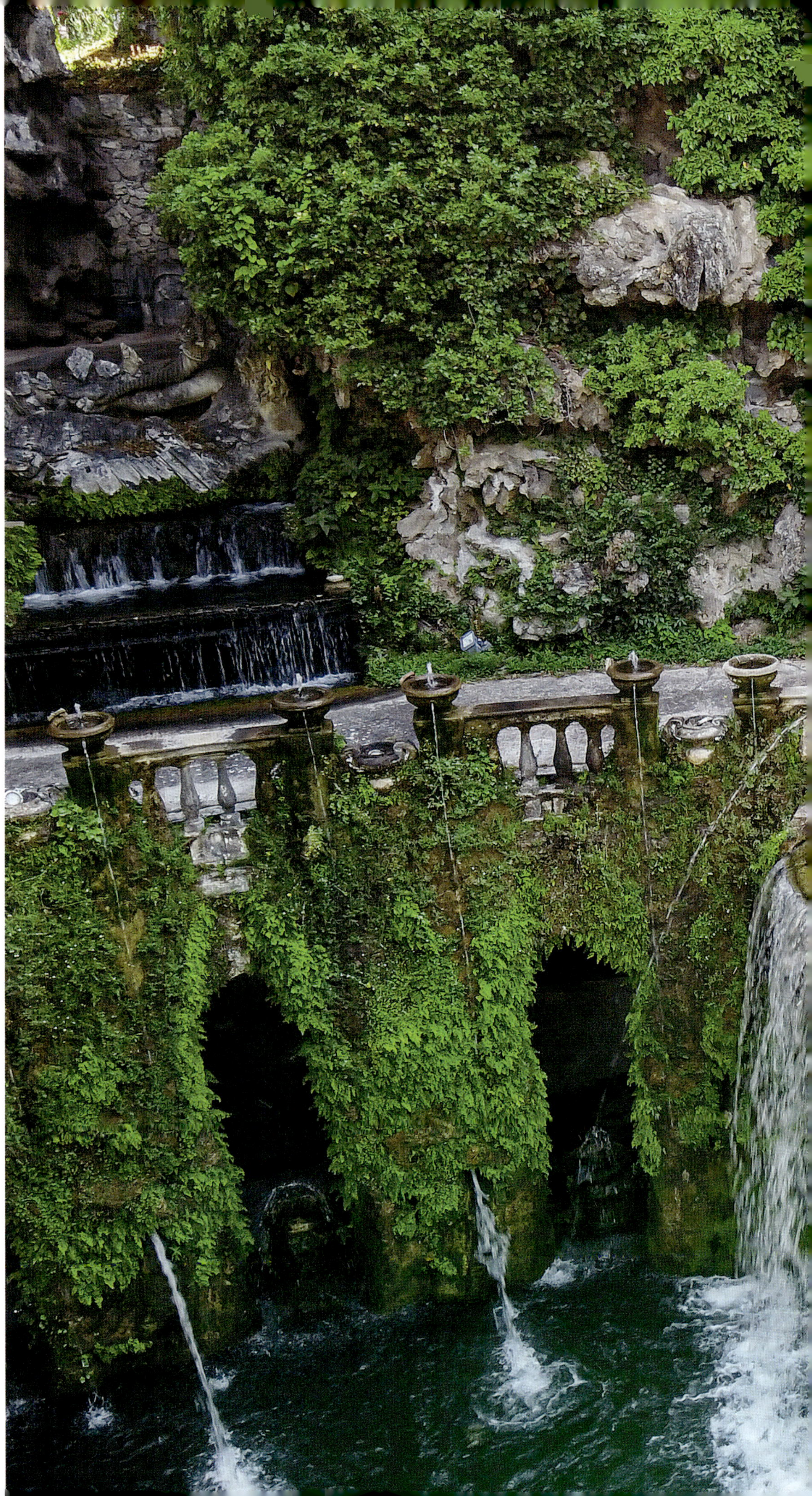

Here and on pages 102–103
Fountain of Ovato, or of Tivoli
Highlights: in the centre is the Tiburtine Sibyl or Albunea, with the baby Melicerte, son of the nymph Ino, symbolising the River Albuneo, made by Gillis van den Vliete, while on either side, within niches, two statues of river gods, by Giovanni Malanca, represent the rivers Aniene and Erculaneo (1566).

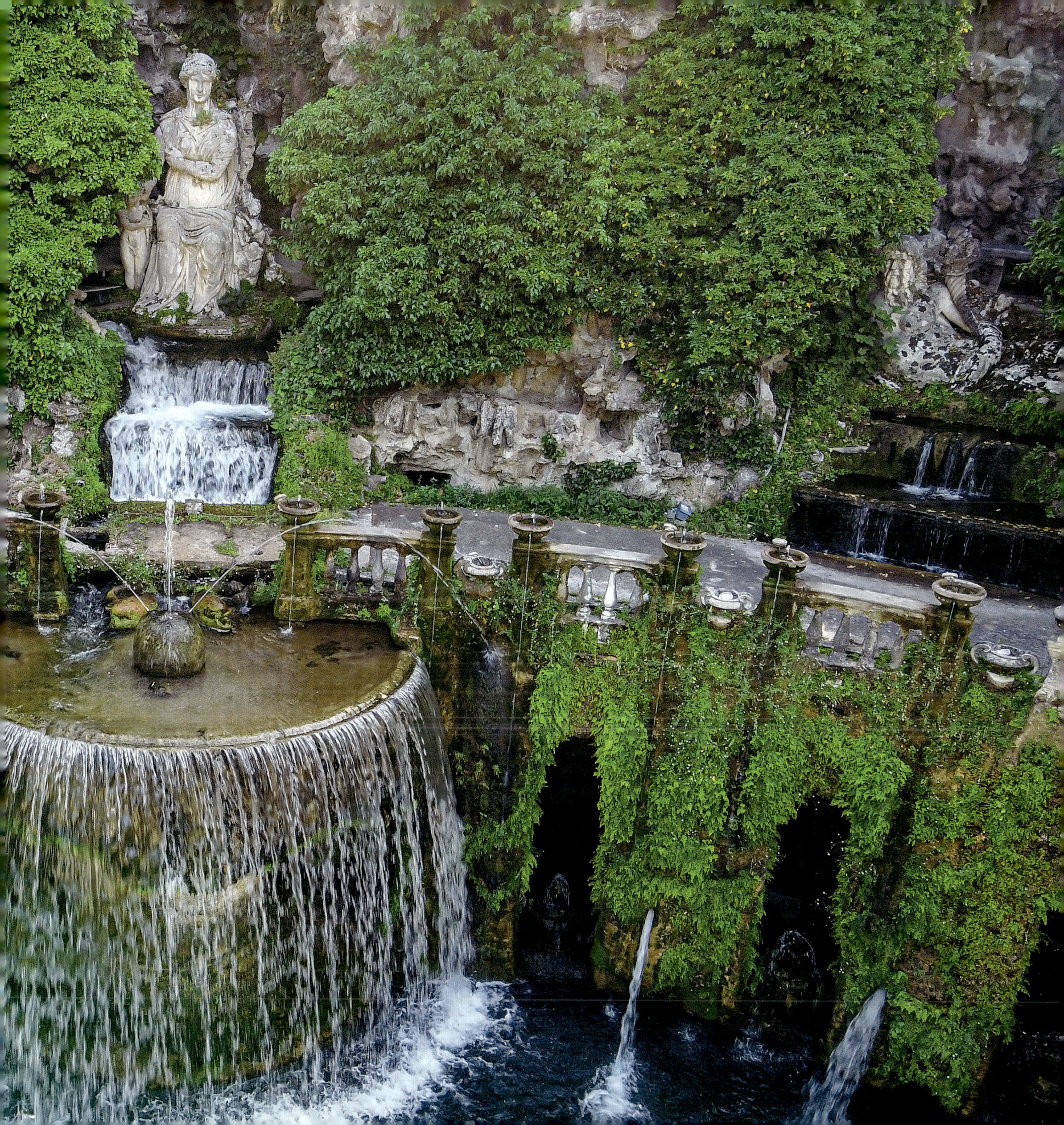

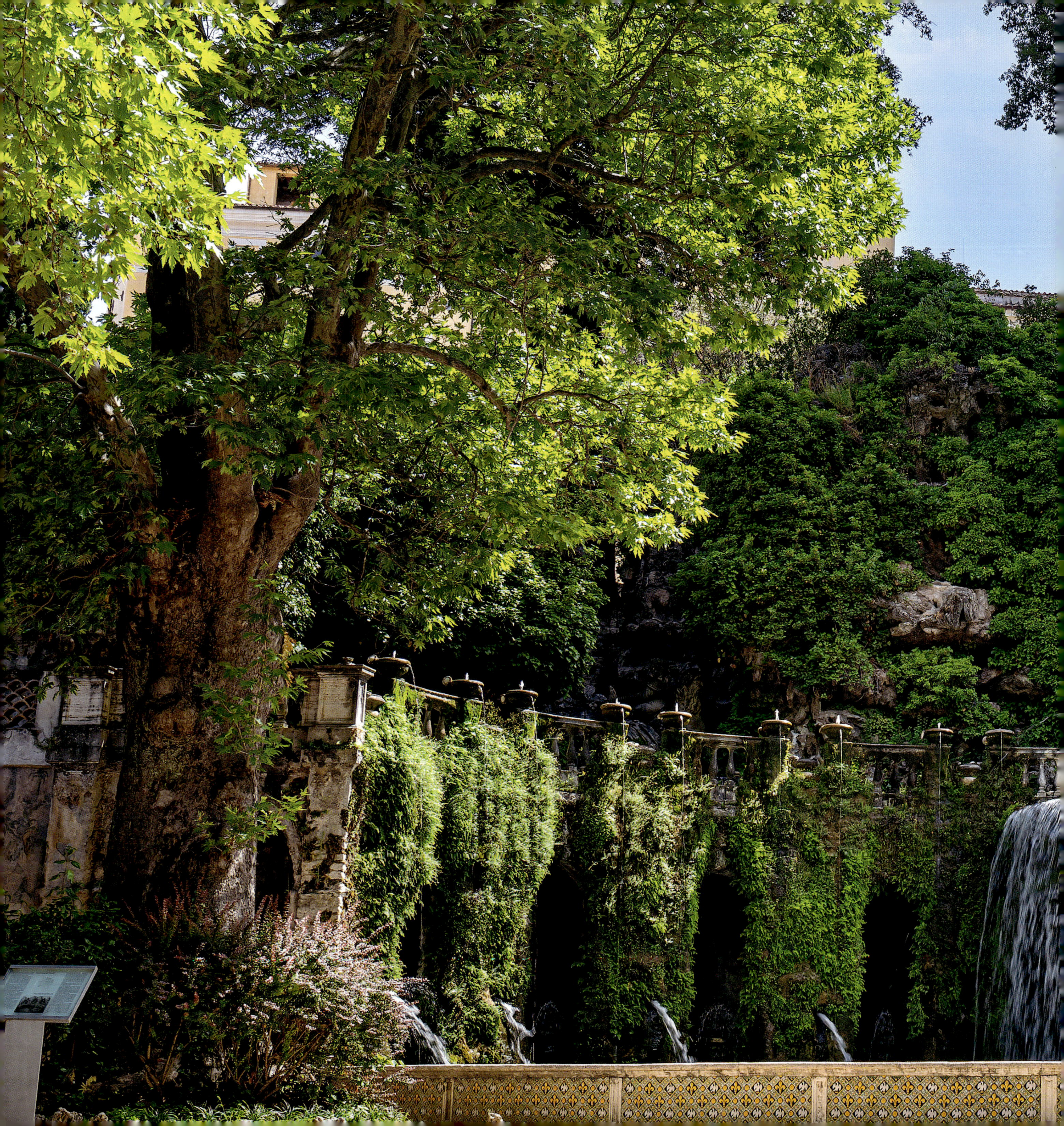

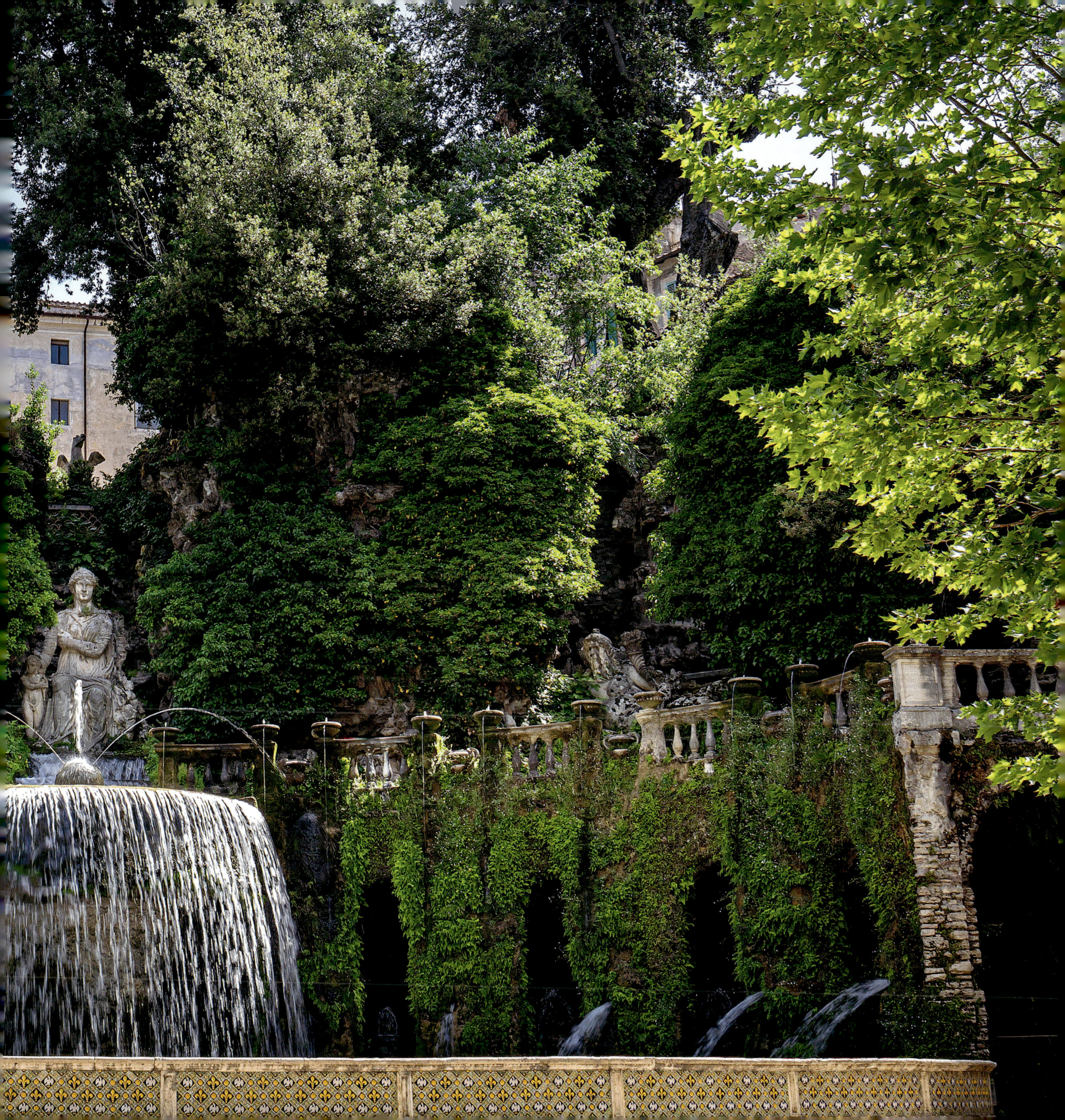

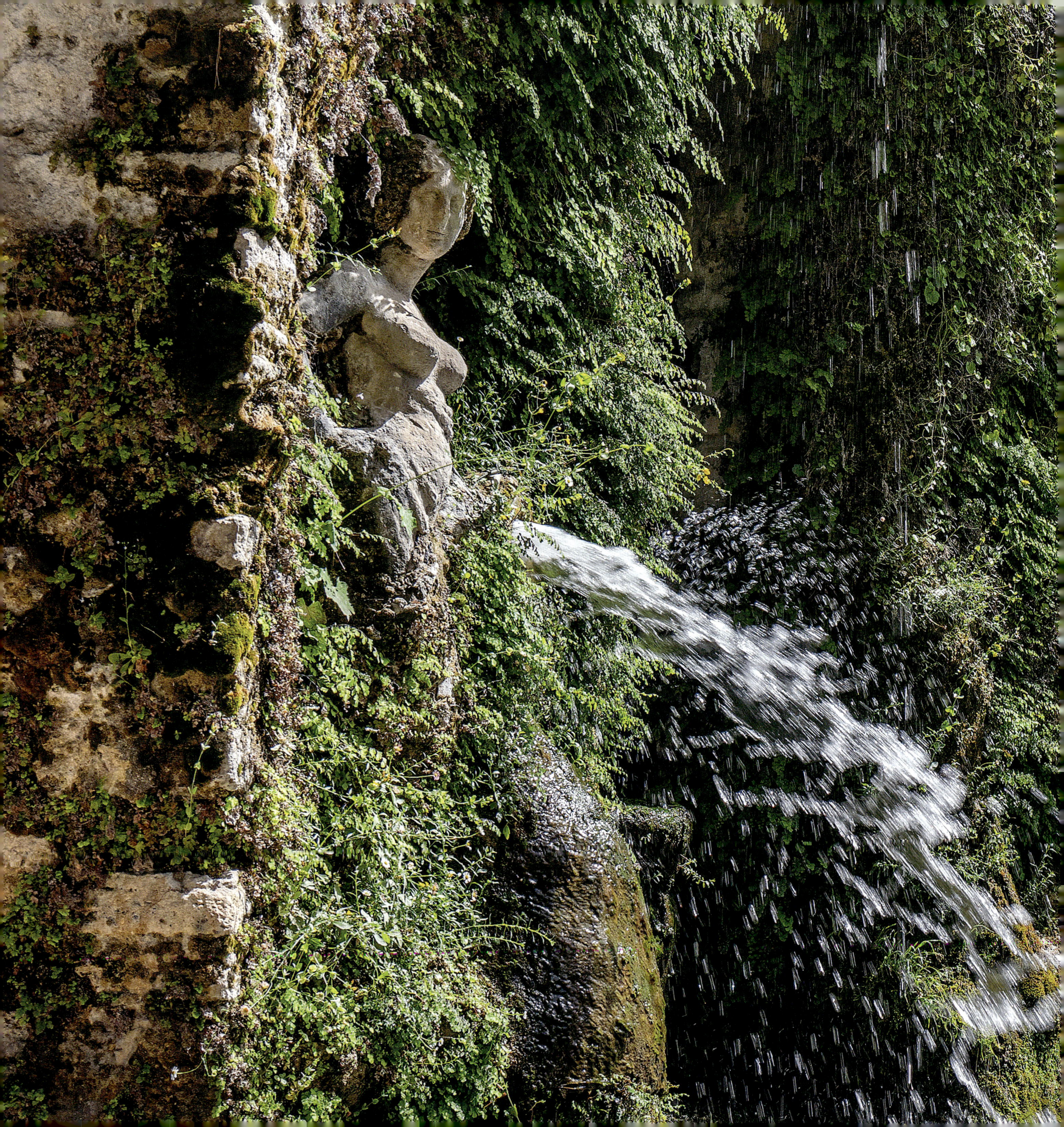

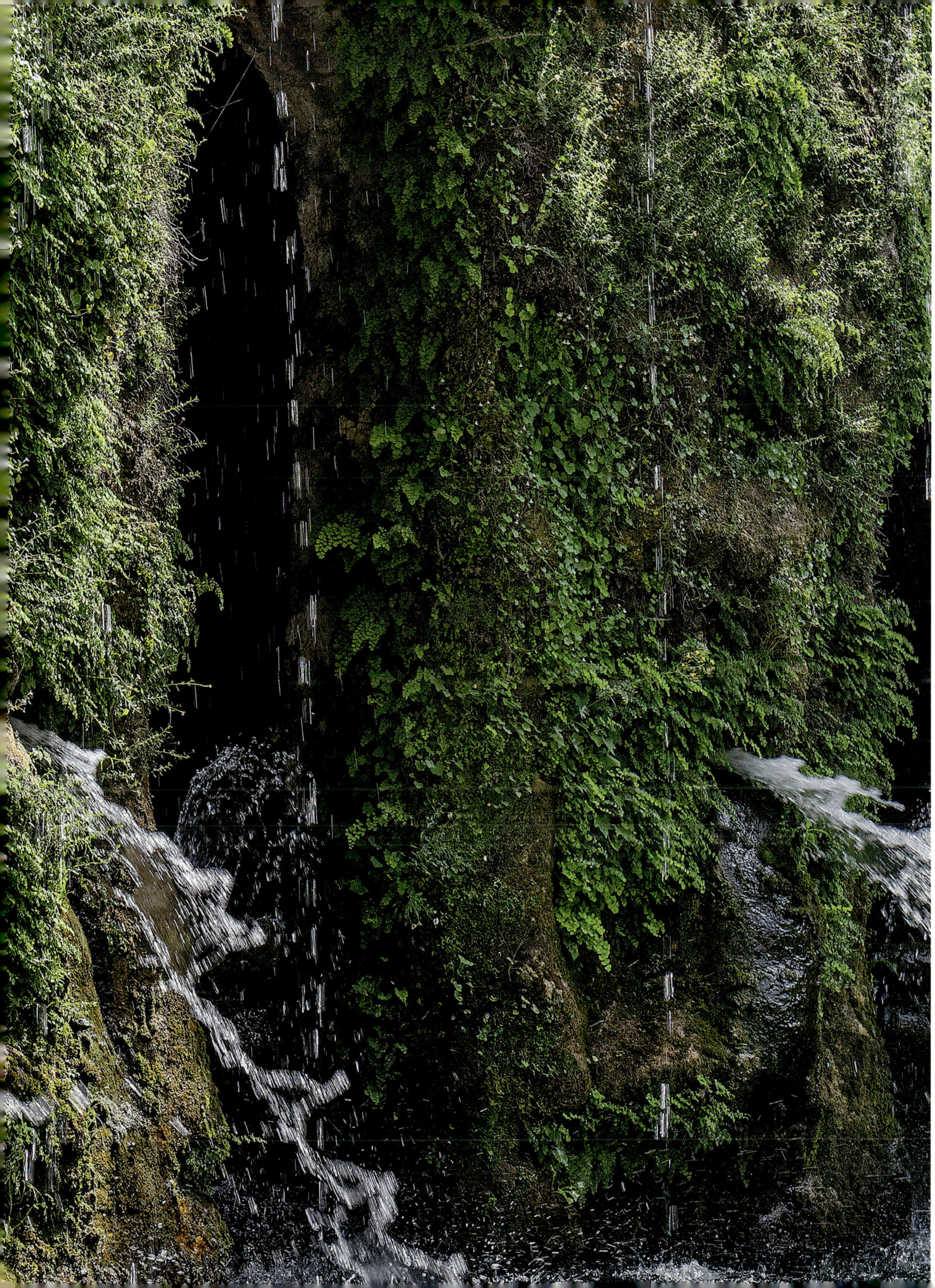

Fountain of Ovato, or of Tivoli
One of ten Nereids pouring water from vases (by Giovanni Battista della Porta to a design by Pirro Ligorio), placed in the niches of a curvilinear nymphaeum, 1567.

by Ligorio and crafted from stuccoed peperino by Giovanni Battista della Porta in 1567. It overlooks a large oval basin with a majolica-covered parapet adorned with Este symbols (largely lost and replaced), created by Bernardino Gentili of Aversa. Behind this section lies a cryptoporticus (currently undergoing restoration) with numerous references to antiquity. In the seventeenth century, the local historian Zappi noted that it was customary to "send women there to make sport of them and have them wetted by many jets of water made with great artifice".

In the upper area, enclosed by a balustrade with water-spouting vases (rebuilt in 1926), there are three grottoes from which streams of water flow. Inside these grottoes, sculptural groups represent the rivers Erculaneo and Anio, sculpted by Giovanni Malanca in 1566, In the centre stands the statue of the Sibyl Albunea with her son Melicerte, by van den Vliete. This overall composition celebrates the Tiburtine territory, emphasising its characteristic elements: rivers, rocks, the Aniene waterfall, and the famous Tiburtine Sibyl. Further upstream, there's a rocky backdrop, artificially constructed from a material known as *pietra spugnola*. Vegetation clings to its rustic heights, forming a crescent-shaped theatre, concealing the original terracotta fistulas, which used to allow water to drip from the rocks. Although the theme of 'sweating rocks' became popular in the seventeenth century and is also found in the opposite part of the garden, today embodied in the Mete, the most spectacular display remains in the centre of the exedra, where a roaring veil-like waterfall overflows from a large crater, with arched jets rising from the top.

Pirro Ligorio's intricate iconological programme, linking the Fountains of Tivoli and Pegasus to the Rometta Fountain on the opposite side, symbolises the desire to recreate metaphorically the physical nature of the local territory while associating it with the grandeur of Rome. This representation is experienced as one progresses along the Avenue of a Hundred Fountains, after catching a glimpse of the Venus Grotto.

Venus Grotto

Situated on the southeast side of the enclosure and built between 1565 and 1568 to Ligorio's design, the Venus Grotto served as a place for meetings and conversations during summer afternoons, as it "offers coolness due to its natural conformation, and cheers with the roar of the external fountain and the murmur of the internal one" (Foglietta). The typology can be traced back to that of the nymphaea, and the entire architectural layout draws inspiration from Roman models. Beyond the walled enclosure, adorned with a celebratory ode by d'Annunzio, one enters the Avenue of the Hundred Fountains.

Hundred Fountains

The scenographic avenue, one of the garden's most spectacular inventions, is 130 metres long. It features a long rectangular basin, followed by three rows of channels designed to follow the terrain's slope. Each channel is equipped with approximately three hundred jets and spouts. Built by 1571 to Ligorio's design, only the general layout of the fountain remains. Historical documents show that the upper channel was crowned by twenty-two small boats interspersed with seventy-two terracotta vases which, arranged in groups of three, held a pomegranate, bitter orange and lemon plants. Although Cardinal Alessandro had already replaced some of the vases with sculptures of Este eagles, the rest of the vases were later removed and the spaces were filled with obelisk-shaped decorations. Very little is visible today of this apparatus due to the silty deposits that, caused by the current condition of the waters of the Aniene river, lead to the formation of encrustations and the constant growth of vegetation. Continuous restoration and cleaning efforts are ineffective against these challenges. Even less remains of the decoration that originally adorned the vertical surfaces of the canals and the basin. These decorations have deteriorated over time due to the material used, namely stucco. The frames featured herms, volutes, and rosettes within which were panels depicting bas-relief scenes inspired by Ovid's *Metamorphoses*.

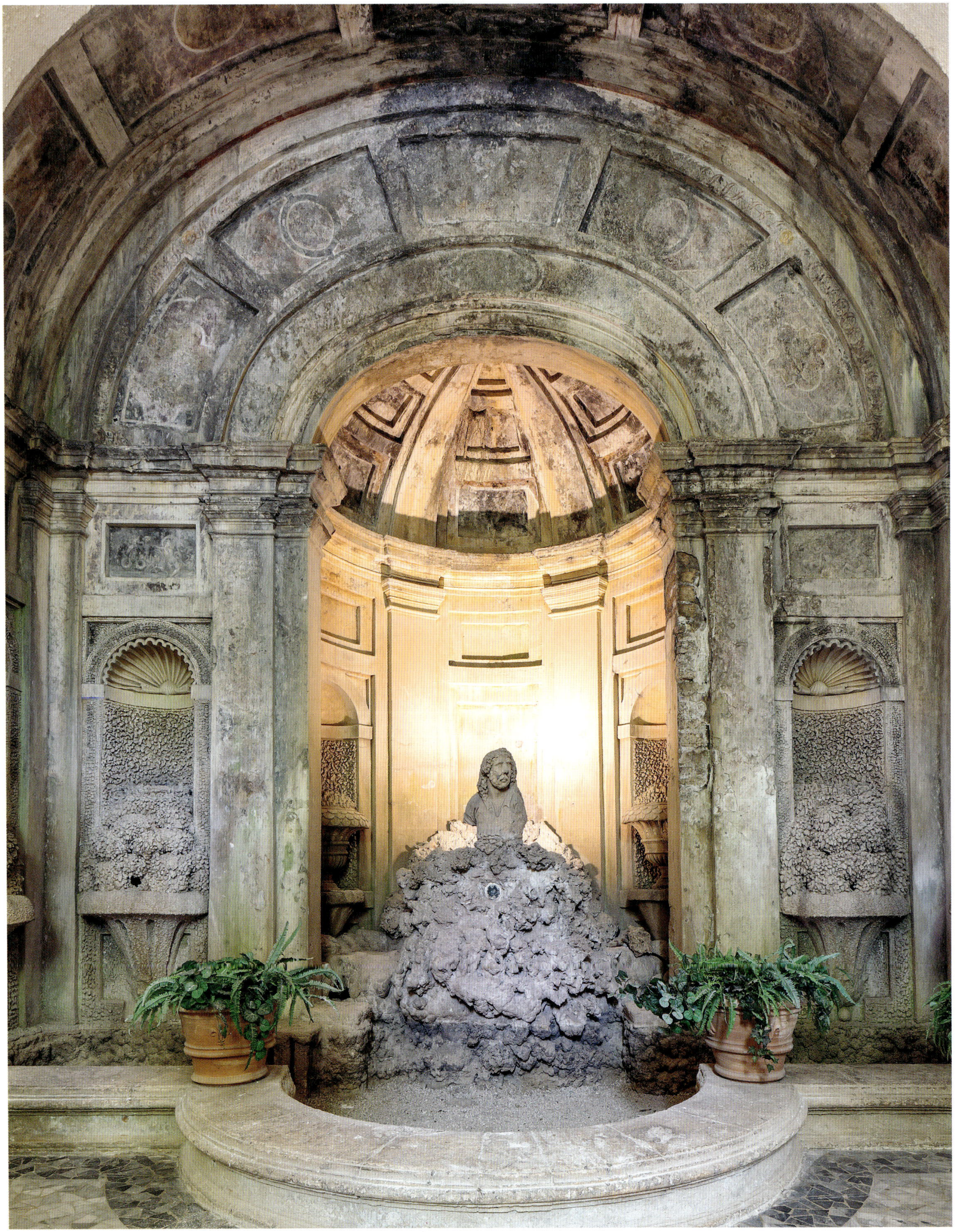

Pirro Ligorio, Venus Grotto, 1565–1568

Page 109
Pirro Ligorio, Hundred Fountains (detail), 1565–1571
Twenty-two small ships were originally positioned on the top, simulating navigation towards Rome and interspersed with vases of limoncelli, pomegranates and melangoli. In 1622, Cardinal Alessandro inserted Este eagles in place of some vases, which were then permanently removed by Francesco II d'Este, who replaced them with obelisks flanked by fleurs-de-lis. Water gushes from these crowning elements of various shapes, and flows down into the canals below through gargoyles and mouths of masks.

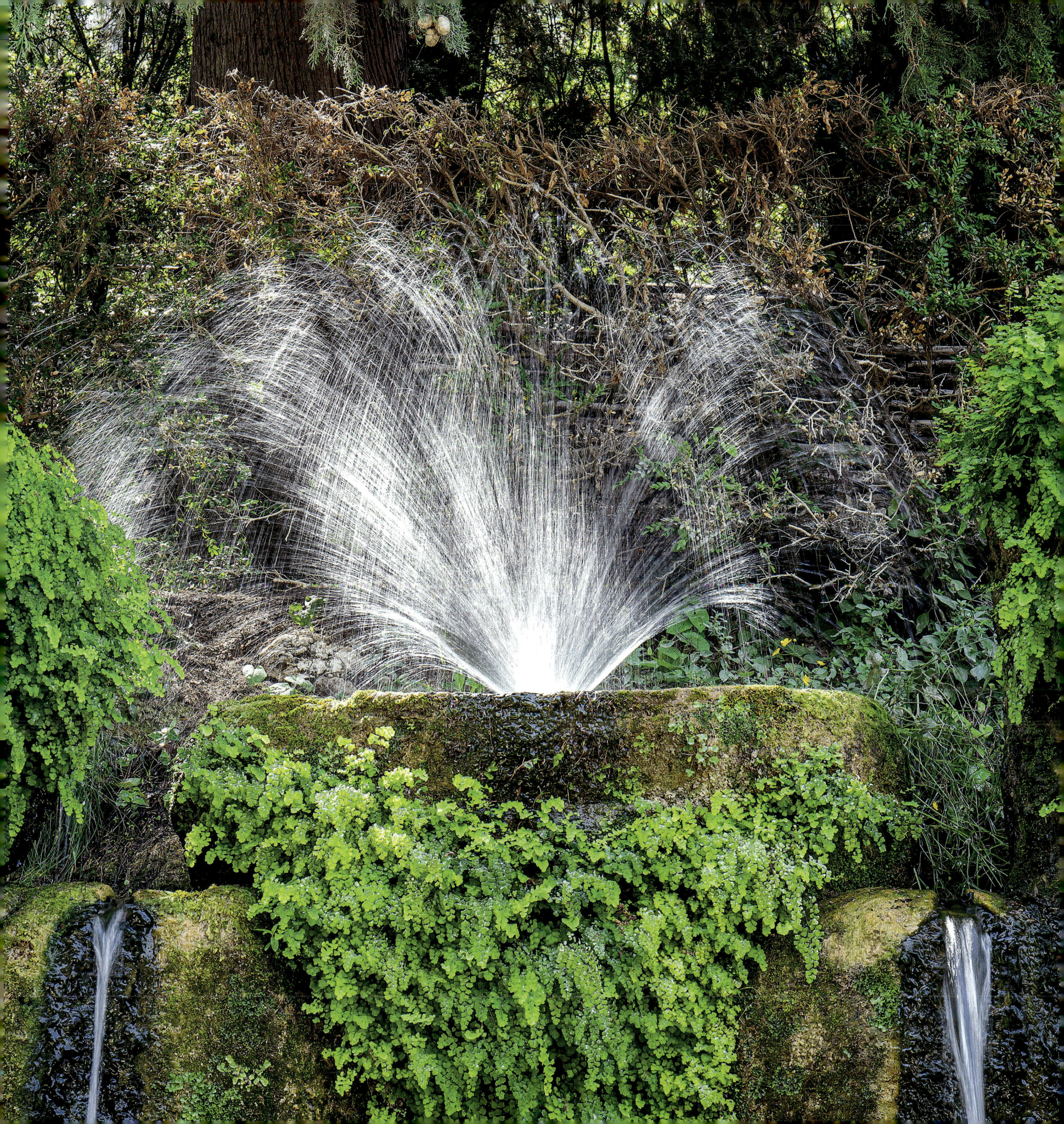

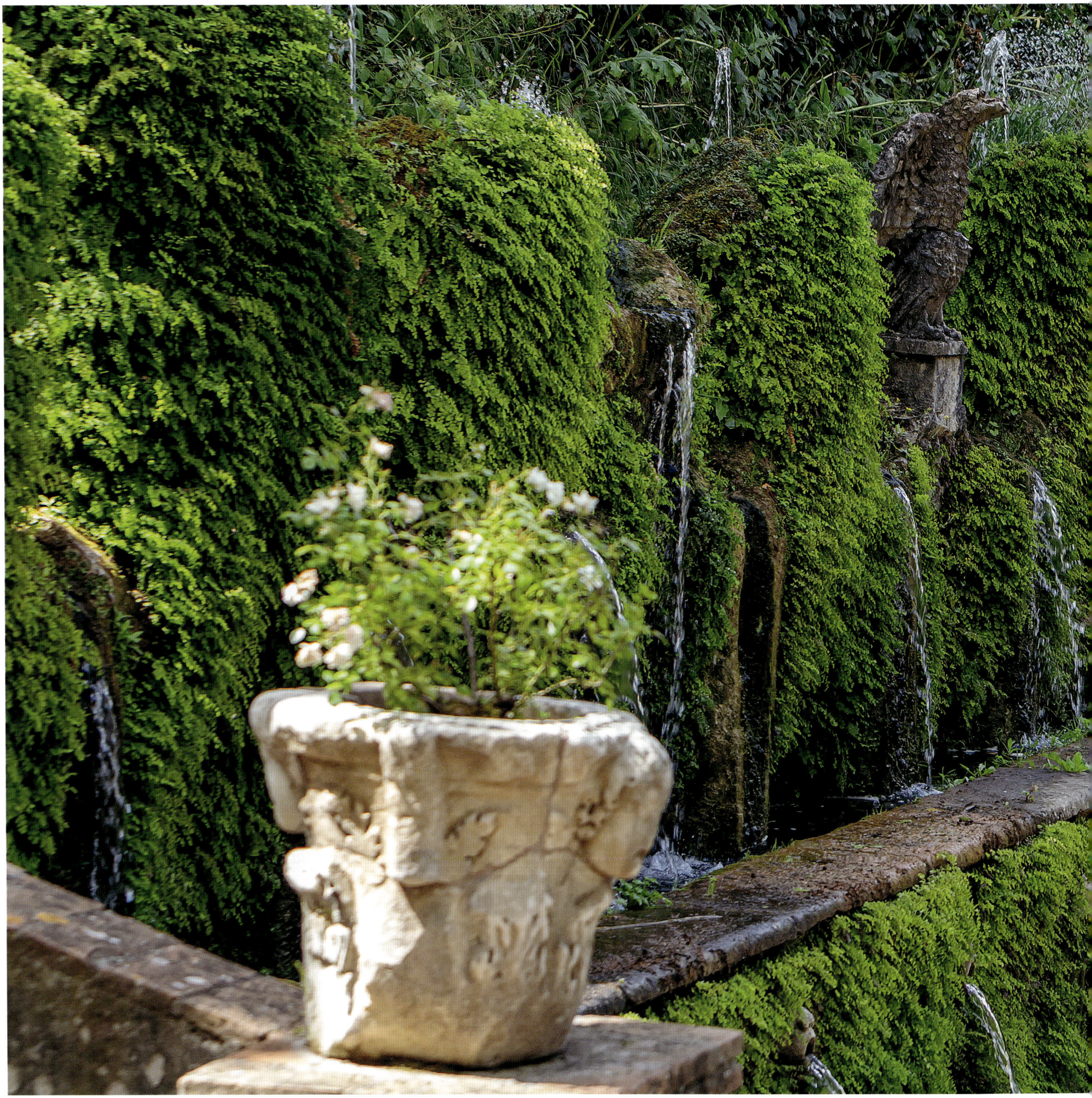

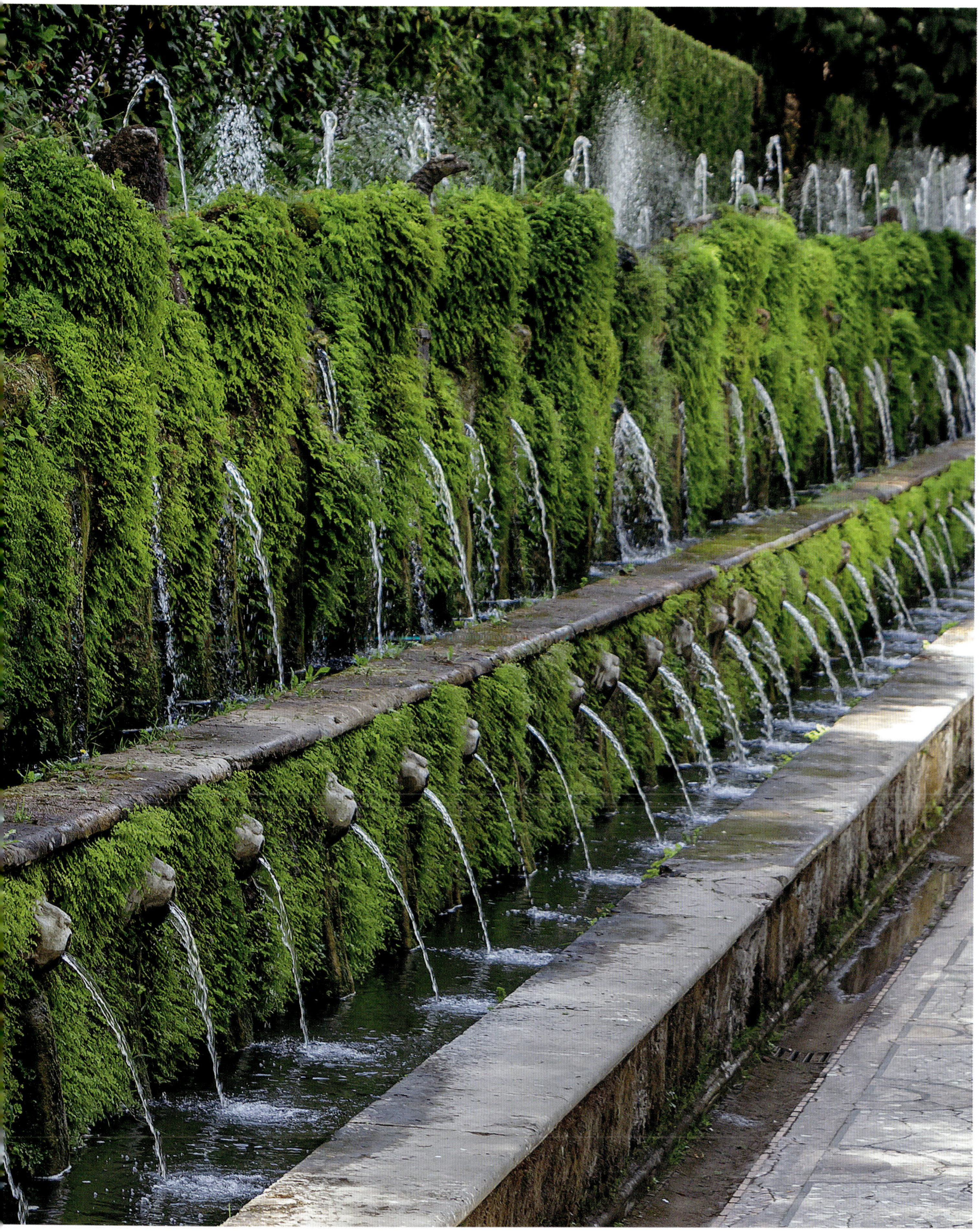

The Hundred Fountains
Here is rendered in a spectacular manner the metaphor of the rivers and aqueducts from Tivoli, taken up in several fountains by Pirro Ligorio. It is an avenue one hundred and thirty metres long that runs transversally through the garden to the Rometta Fountain, towards which the waters flow.
The avenue is flanked on the uphill side by a basin containing almost three hundred jets arranged in three superimposed canals.

Page 112
Avenue of the Hundred Fountains with in the background the Rometta Fountain

Page 113
Zoomorphic mask of the Hundred Fountains

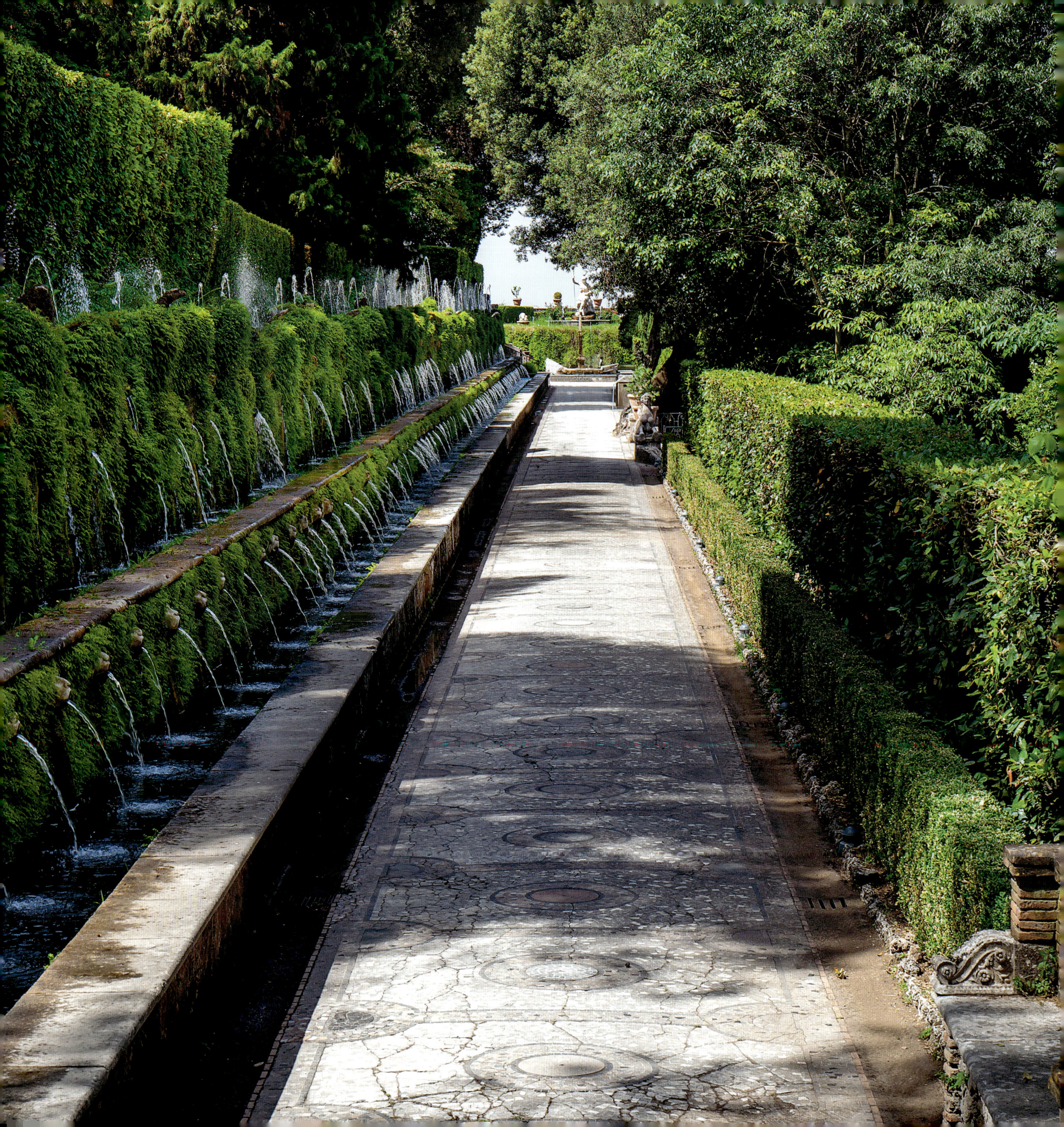

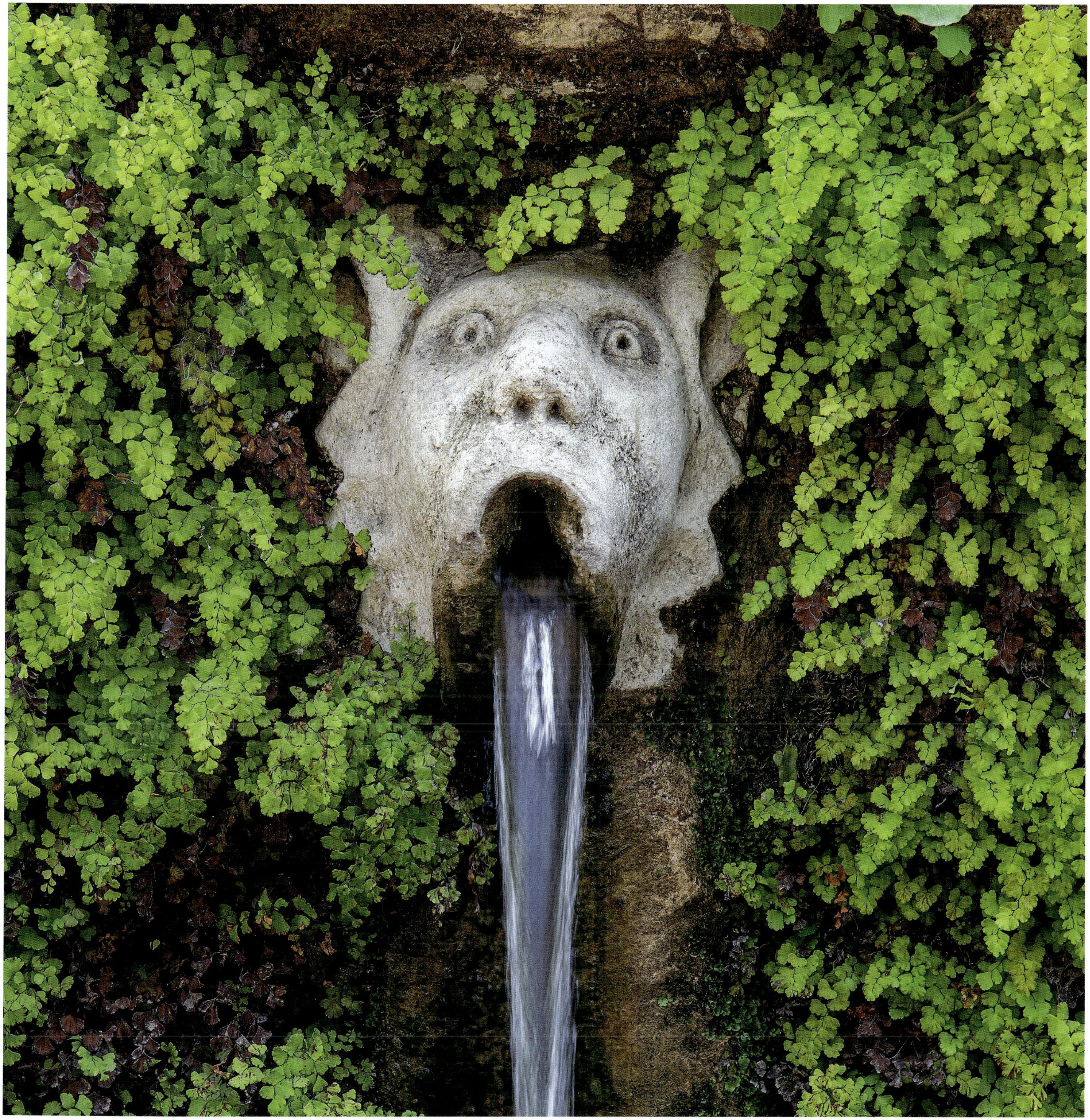

The espalier of the intermediate channel, adorned with zoomorphic masks – certainly taken from ancient models, some of which are present in Hadrian's Villa – contained mosaic decorations depicting fish, birds, and sea monsters, reflecting common ancient iconography. In a substantial restoration carried out in 1930 by director Attilio Rossi, missing eagles and fleurs-de-lis were replaced with cement casts. The avenue, originally made of beaten earth, was paved with marble recovered from archaeological excavations. While this restoration was quite invasive, it heightened the evocative impact of the Viale Grande, celebrated by d'Annunzio. Numerous documentaries, filmed by the Istituto Luce from 1930 onwards, including those featuring visits by foreign dignitaries and a visit by Rita Hayworth filmed in 1950, highlight the grandeur of this restoration. Additionally, scenes from Woody Allen's 2012 film *To Rome with Love* are set in this remarkable location.

Rometta Fountain

The Avenue of the Hundred Fountains leads to the antiquarian area of the villa, in which the antique was remodelled as a theatrical backdrop, with Rome as the iconographic subject and the real background on the horizon.

Designed by Pirro Ligorio, and built by Curzio Maccarone between 1567 and 1570, the water theatre sits on a large terrace, supported by substructures. According to a tale passed on by Audébert (1576), in response to Pius V's ban on the construction in Rome of a palace as grand as a castle, Ippolito II replied that if he was not allowed a castle in Rome, he would have Rome in his villa. Due to structural issues, the present configuration results from the demolition of much of the architectural backdrop in 1850. The original backdrop created an enclosed space, similar to the Tivoli Fountain, and allowed for a duplicate image of Rome on the horizon. It offered a close-up, miniature *Urbe* that fascinated visitors. The sequence of small buildings, arranged on a broad exedra, ideally represented ancient Rome as reconstructed by Ligorio. This perspective was the foun-

Pirro Ligorio, Luigi Maccarone,
Rometta Fountain (detail), 1567–1570
Pierre de la Motte's sculpture of Rome Victorious, *is based on a design by Ligorio.*

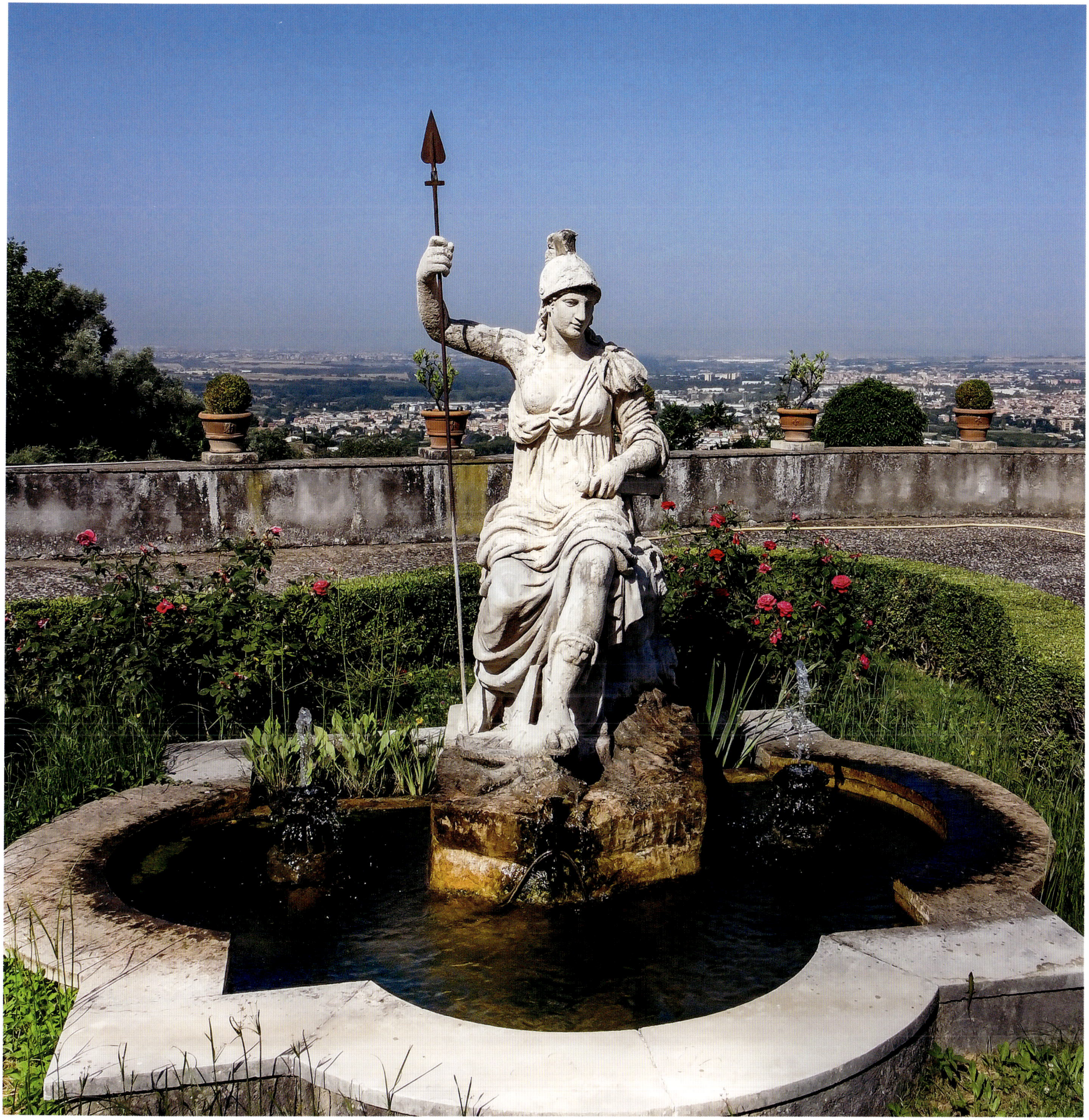

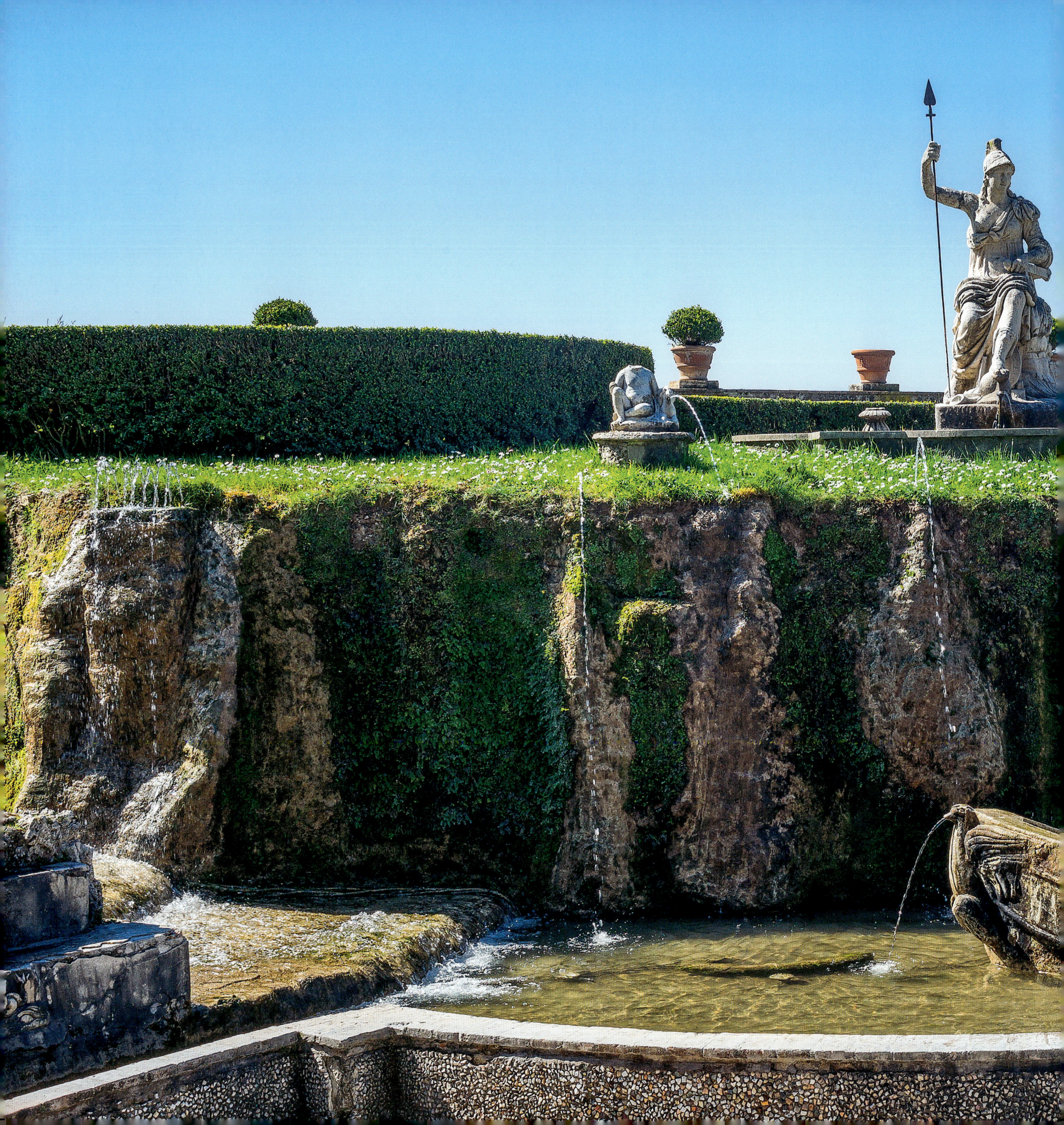

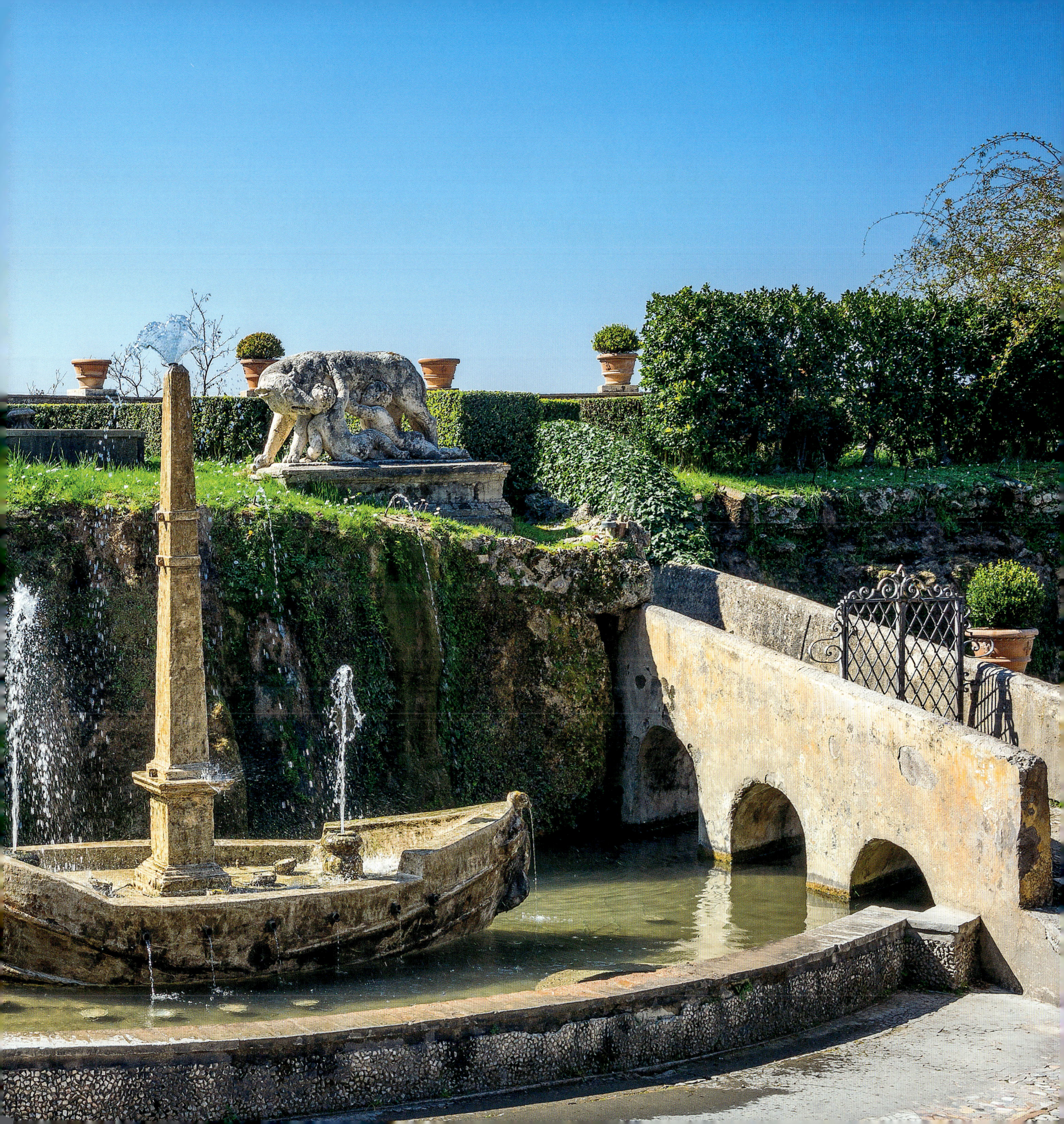

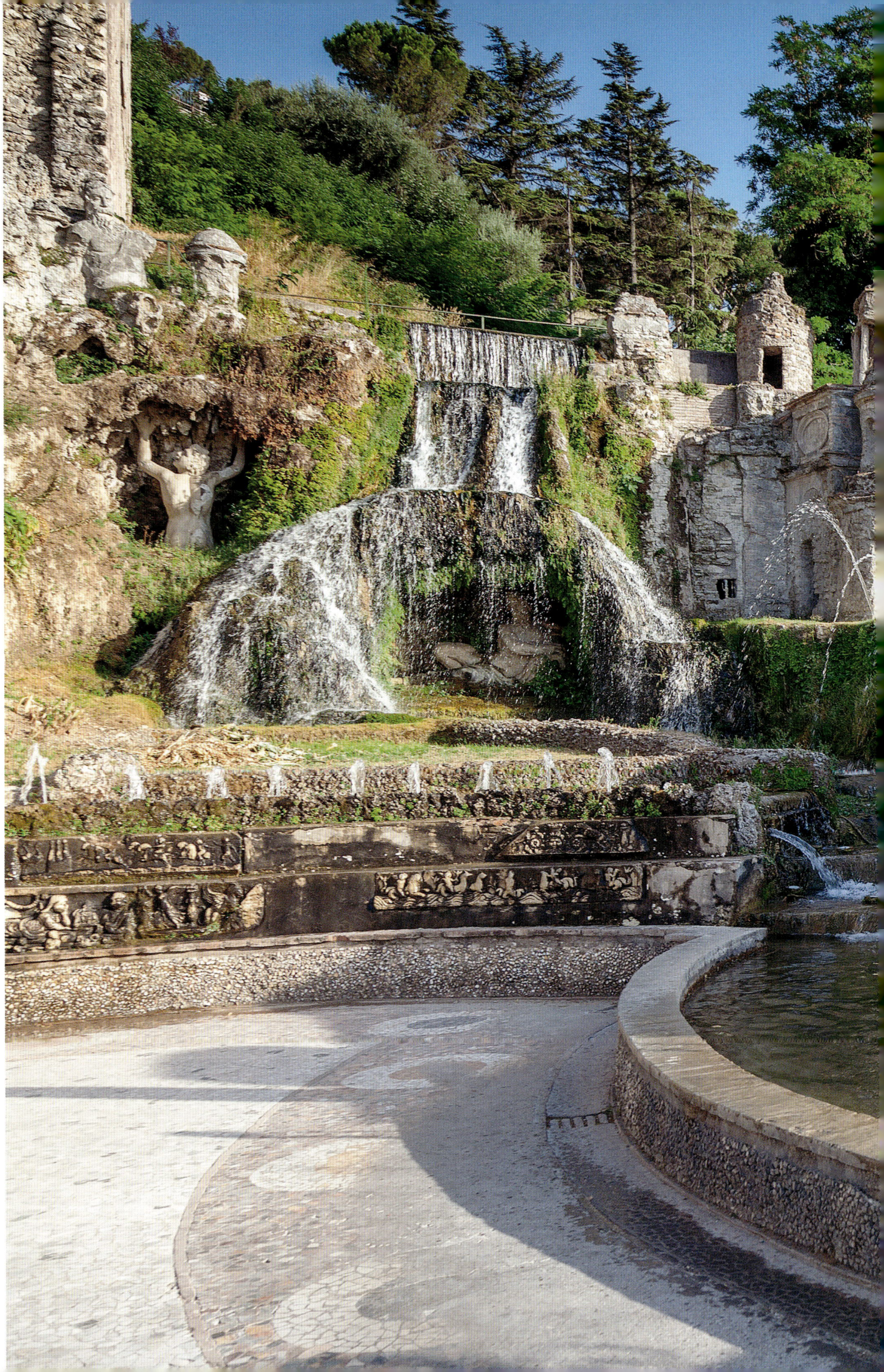

Here and on pages 116–117
Views of the Rometta Fountain
Originally, the fountain was complemented by many other reproductions of monuments of ancient Rome (Arch of Titus, Arch of Septimius Severus, Arch of Constantine, Trajan's Column, Pantheon, Colosseum) sculpted by Pierre de la Motte to a design by Pirro Ligorio, of which no trace remains, except in Giovanni Battista Venturini's engraving dated 1691.

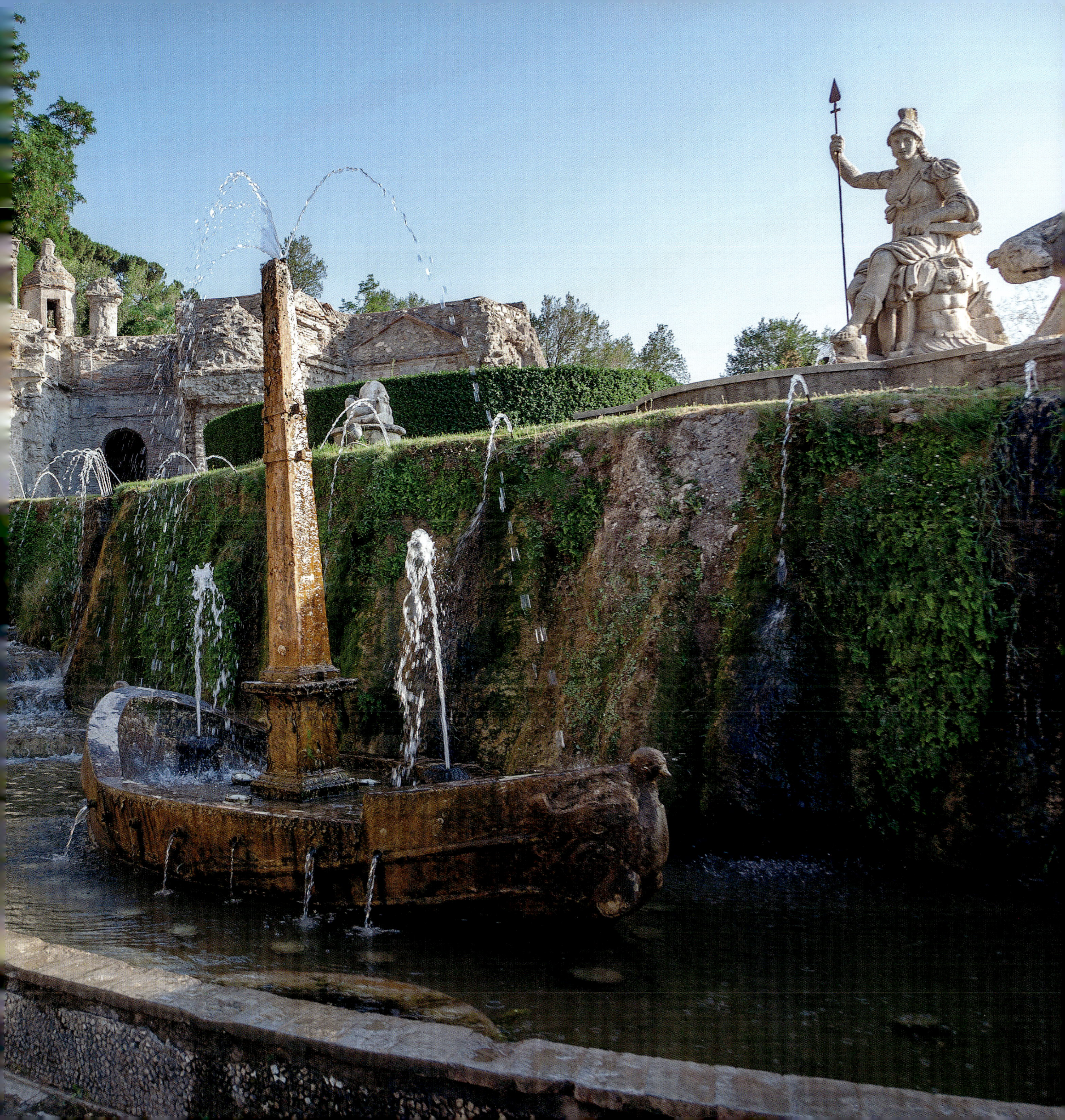

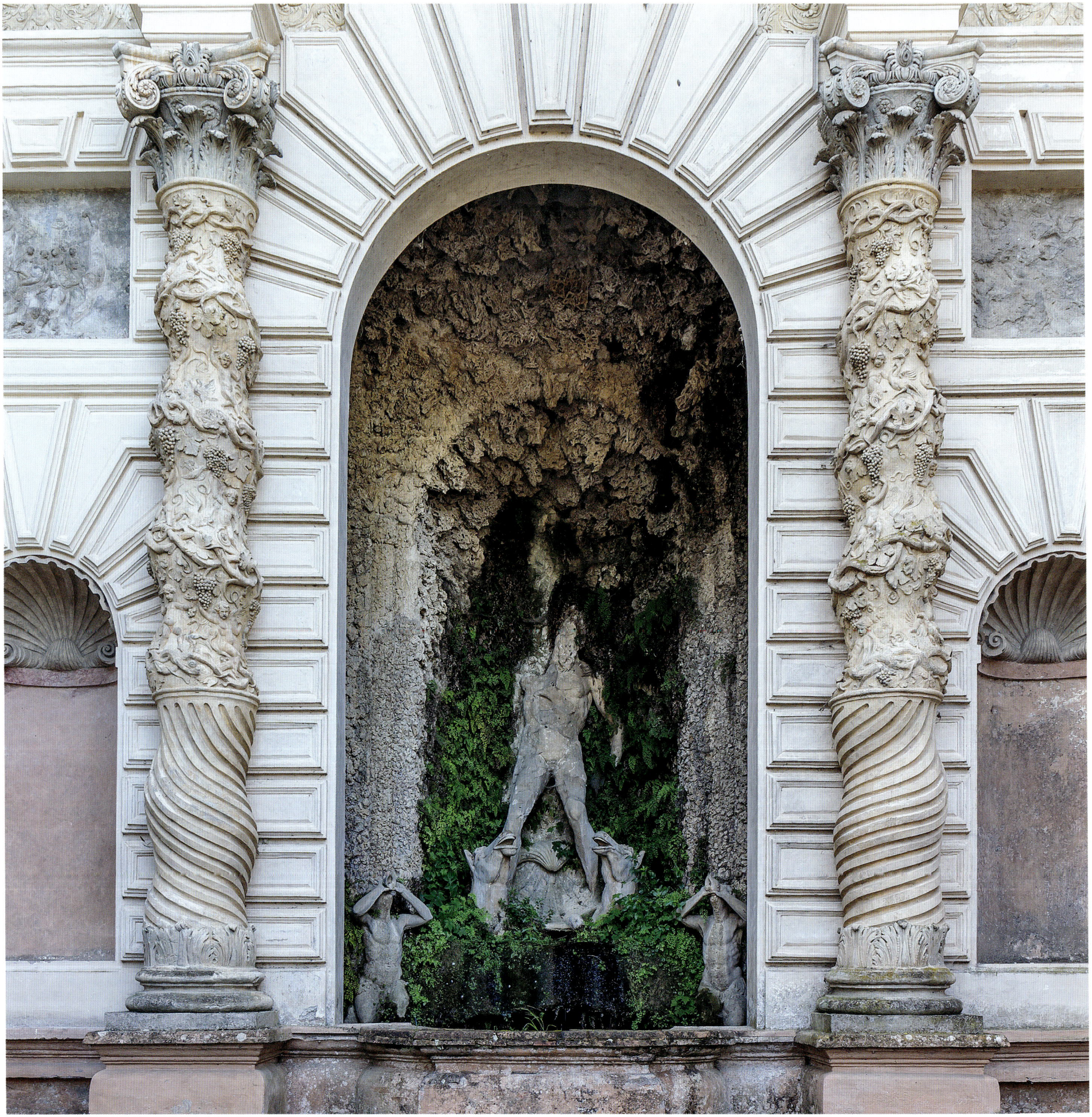

dation of the sculptural perspective and included representations of the Pantheon, the Capitol, the Mausoleum of Augustus, and Hadrian's Mausoleum, according to historical accounts. Comprising seven parts, mirroring the hills of Rome, with key monuments connected by portals symbolising the city's gates, triumphal arches, and aqueducts, this composition directly references the *scaenae frons* of the Roman theatre system. This style was later revived by Palladio for the construction of the theatre in Vicenza. Sculptures representing 'victorious Rome', the *Sibyl* of the Tivoli Fountain, the *She-wolf suckling the twins* (completed in 1611), and the ship, still visible today, condensed the image of the Eternal City into a few symbolic representations.

At the beginning of the seventeenth century, the slope to the north was modelled to resemble a mountain, evocative of Tivoli's morphology, with inspiration drawn from the Apennines (the sculpture within the highest grotto) and watercourses descending from those lands into the Tiber (the sculpture in the second grotto). These waters feed the reservoir at the lower level, in the middle of which the little ship with the obelisk alludes to the Tiber Island. While a group of desecrating volcanic tuff statues were removed at the end of the seventeenth century, some of the famous fragments and friezes of heroes and marine divinities remain in the base of the basin, and were taken from the area of the Maritime Theatre of Hadrian's Villa, some pieces of which are in the Museo Nazionale Romano, and some of which were replaced with copies after the originals were moved to the *Mouseia* of the imperial villa.

Fountain of Proserpine

This nymphaeum, designed in the form of a triumphal arch with twisted columns, was completed in 1570 by Giovanni Alberto Galvani. It served to resolve a difference in height in an elegant way. Turning away from this nymphaeum, you'll arrive at the area of the Owl Fountain.

Giovanni Alberto Galvani,
Fountain of Proserpine, 1570
The nymphaeum is in the form of a triumphal arch with tortile columns wrapped in stucco vine shoots. The architect from Ferrara, Galvani, had initially planned the Fountain of the Emperors with statues of Caesar, Augustus, Trajan and Hadrian to decorate the corners, while in the central niche, the fountain is decorated with a fountain of the Emperors, while the central niche was to contain the statue of Arethusa and other nymphs. However, these were never executed and were replaced in the seventeenth century by the stucco group of Pluto abducting Proserpine, on a shell-shaped chariot pulled by horses, while two Sileni play sea harps and two dolphins stir the waters. The figure of Proserpine, together with the two statues in the niches, clearly visible in an engraving by Venturini and dated 1685, have all been lost.

Owl Fountain

This is the first of the three sound fountains completed between 1565 and 1569, that were a great source of interest and attraction in the villa. Designed by Ligorio and brought to life by Giovanni Del Duca and the French fountain-maker Luc Leclerc, the Owl Fountain features a central courtyard enclosed by high walls adorned with arches, niches, and water-spouting basins. The primary facade boasts an elegant niche flanked by Ionic columns and crowned by the Estense eagle. The entire facade is embellished with mosaic decorations echoing Este themes and symbols, while only remnants of the original stucco decor remain within the niche. Of particular interest is the hydraulic artifice, crafted by Leclerc, where twenty painted bronze birds sing until an owl appears. This ingenious device, rooted in the writings of Heron of Alexandria, a scientist from Hadrian's time known for creating 'automata', regrettably, has been only partially restored. It is less debated that French expertise in such devices may have originated from Leonardo da Vinci, who, armed with knowledge of Heron's treatises, created a 'mechanical lion' for François I of France in 1515. Leonardo's designs for the *jardin de rève* for Charles d'Amboise also featured a mechanical bird, which could relate to the automaton created for Agnolo Poliziano's play *Orpheus*. In 1517, contact with French culture sparked new inspiration in Leonardo, who was highly regarded for his production of machines that evoked "amazement and wonder". These machines featured intricate mechanical devices and were later cherished by Ippolito II during his visits to France, prompting him to seek Leclerc's expertise to enhance his garden in Tivoli.

Leaving the Owl Fountain and heading east, a further promenade crosses the garden from west to east, parallel to the main front of the palace to the north. Along this path one intercepts some steps known as the Bollori.

Giovanni Del Duca, Owl Fountain, 1565–1569
Located at the bottom of the Avenue of Dragons, it composes – with the Fountain of Proserpine and Rometta – a spectacular sequence of sloping nymphaea, set on the artificial terraces that structure the garden to the south-west.

Pages 124–125
Detail of the pediment of the Owl Fountain

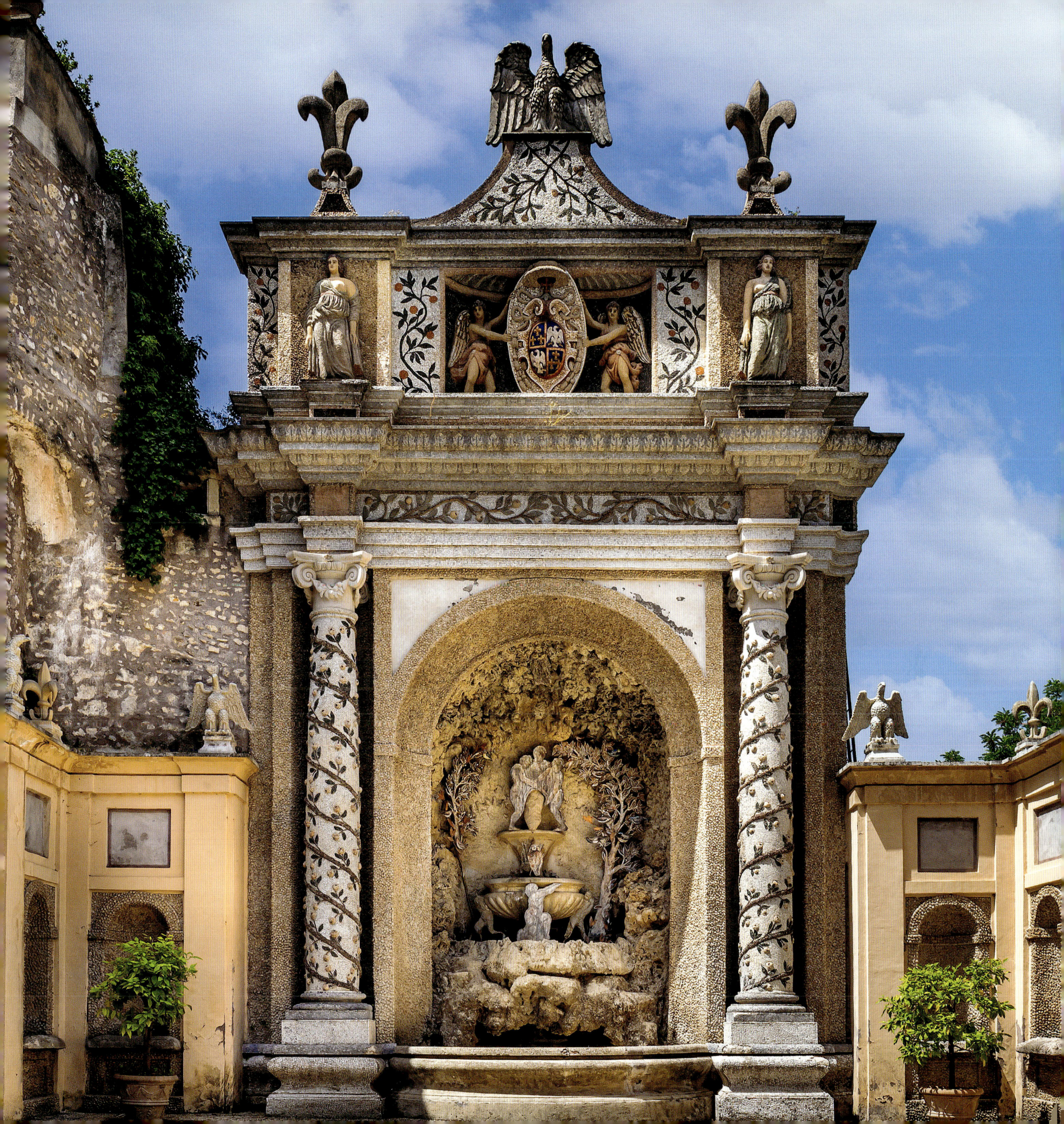

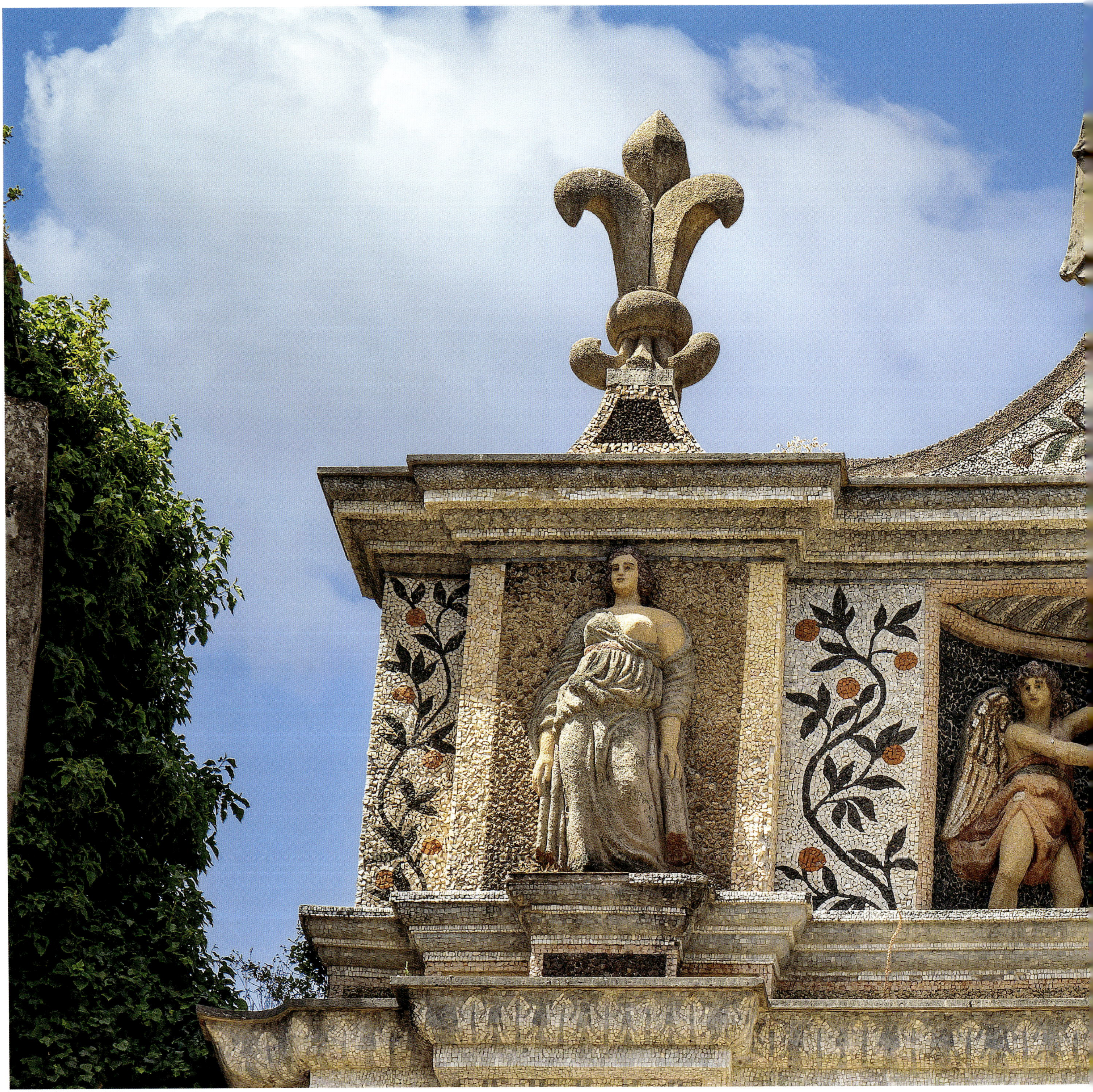

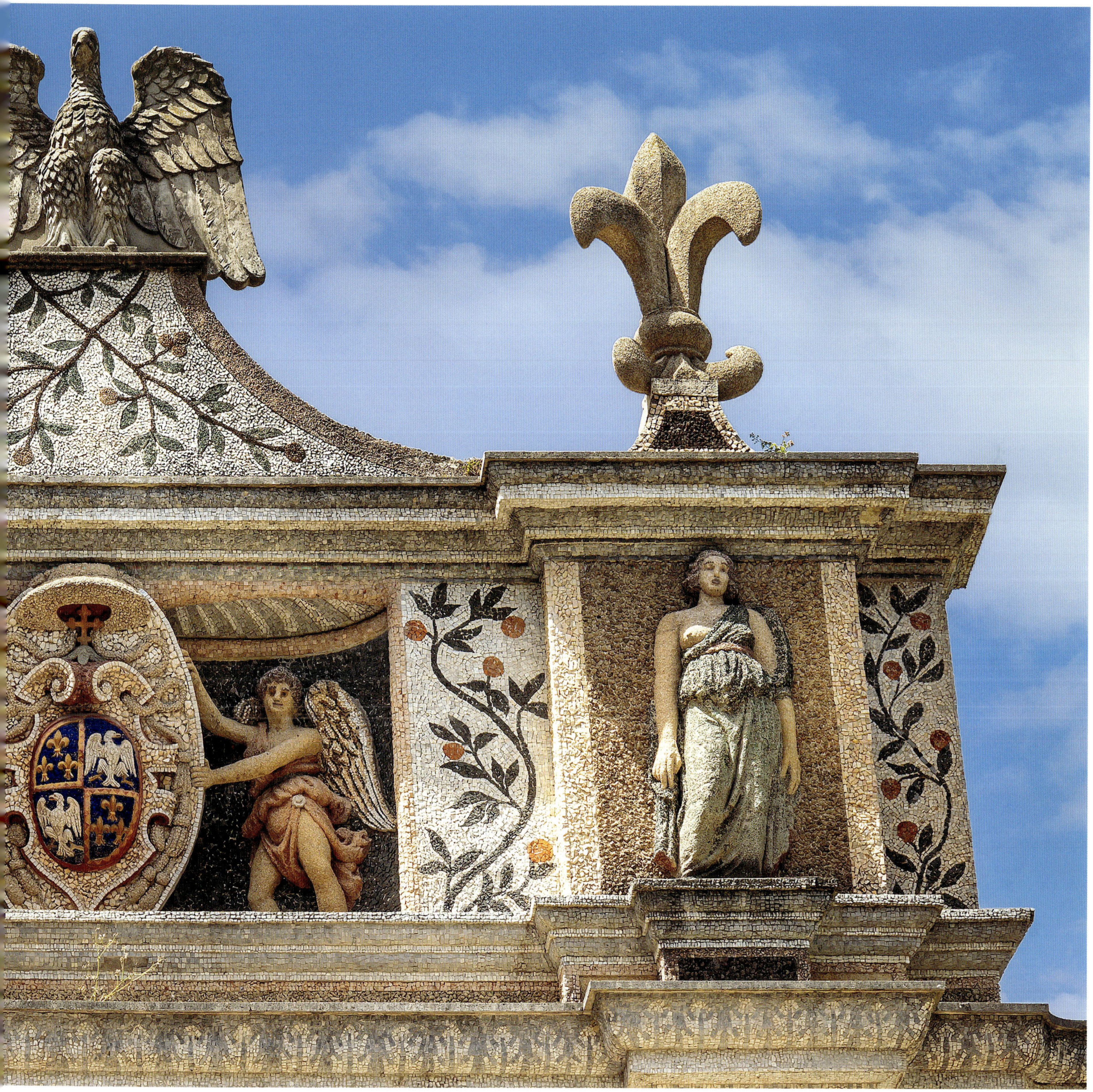

Bollori

The name 'Bollori' ('boilings') originates from the vapour generated by the gushing water, which appears to boil as it pours profusely from parallel structures on either side of the ramps. Constructed by 1753, according to Ligorio's design, these stone features closely resemble the Water Staircase of the Moorish Jardines Altos at the Alhambra in Granada, which began to influence designs after 1492, following the conclusion of the War of Granada. This influence grew following the city's conquest by the Catholic Monarchs, Ferdinand II of Aragon and Isabella of Castile.

Fountain of the Dragons

Positioned in line with the front of the palace and the entrance to the valley, the Fountain of the Dragons, which like the neighbouring Owl Fountain is enriched with sound effects, underwent considerable remodelling from the year of Ippolito II's death and again in the seventeenth century, when it was given the name 'Fontana della girandola' (Fountain of the pinwheel) following the inclusion of special water features. However, Ligorio's original design already made it a marvel in the garden. Sound effects that simulated shots and bursts of firearms were achieved through hydraulic artistry, masterminded by Sienese architect Tommaso Ghinucci, utilising different water pressures. In celebrating Hercules' eleventh labour in the Garden of the Hesperides, this theme was thus closely associated with the exploits of Ippolito II and his household and was represented by a statue of Hercules with club, placed in the central niche (now in the Louvre, while the current sculpture depicting Jupiter is the result of seventeenth-century interventions), as well as statues of gladiators and Mars in the side niches. The fountain features a large basin which, in the centre, shows a group of four winged dragons whose mouths jets of water shoot out and from which the powerful central jet rises. Anecdotal legend has it that this group was set up in a single night, in September 1572, on the occasion of the visit of Pope Boncompagni,

View of the Bollori stairs, 1753

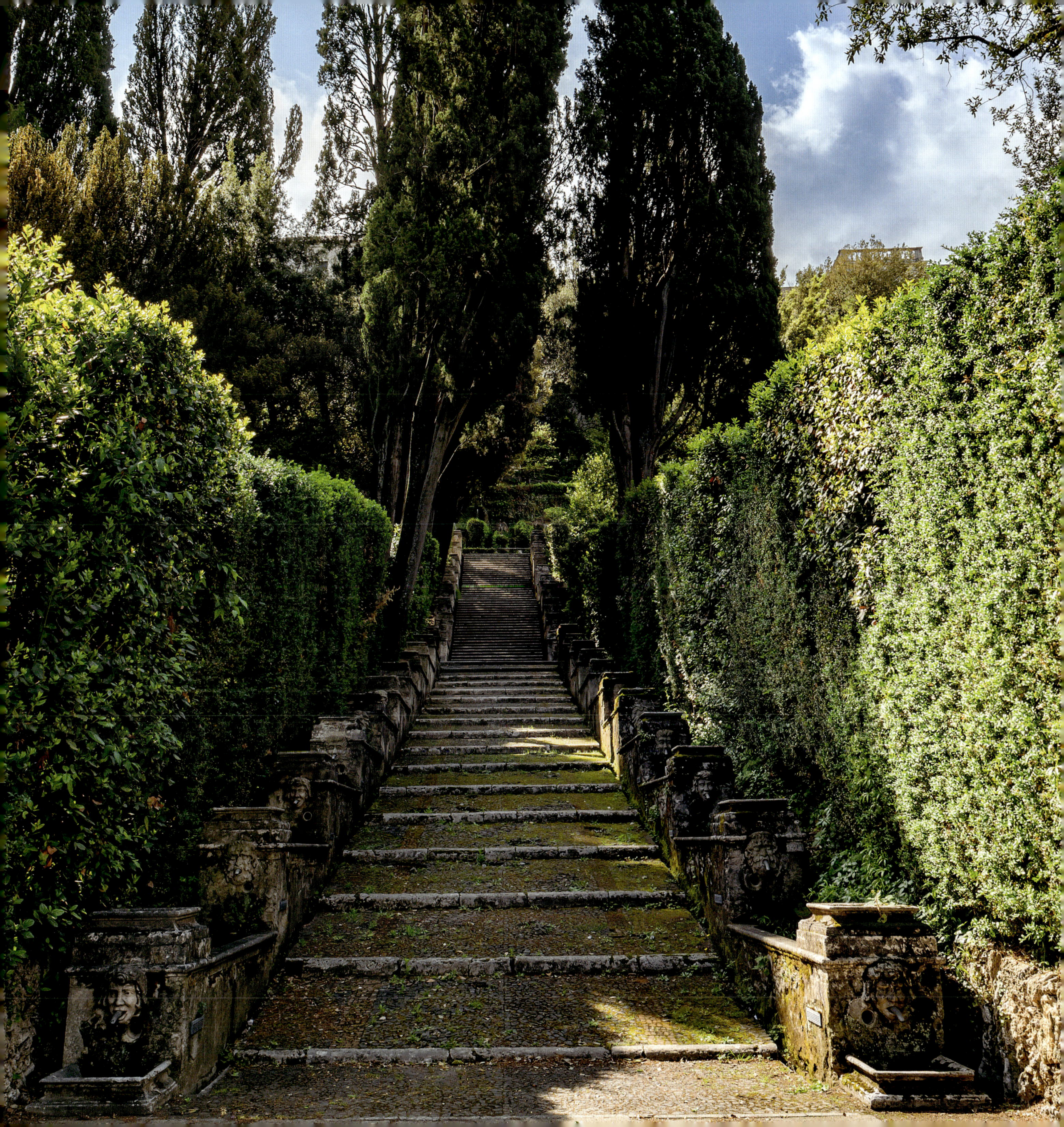

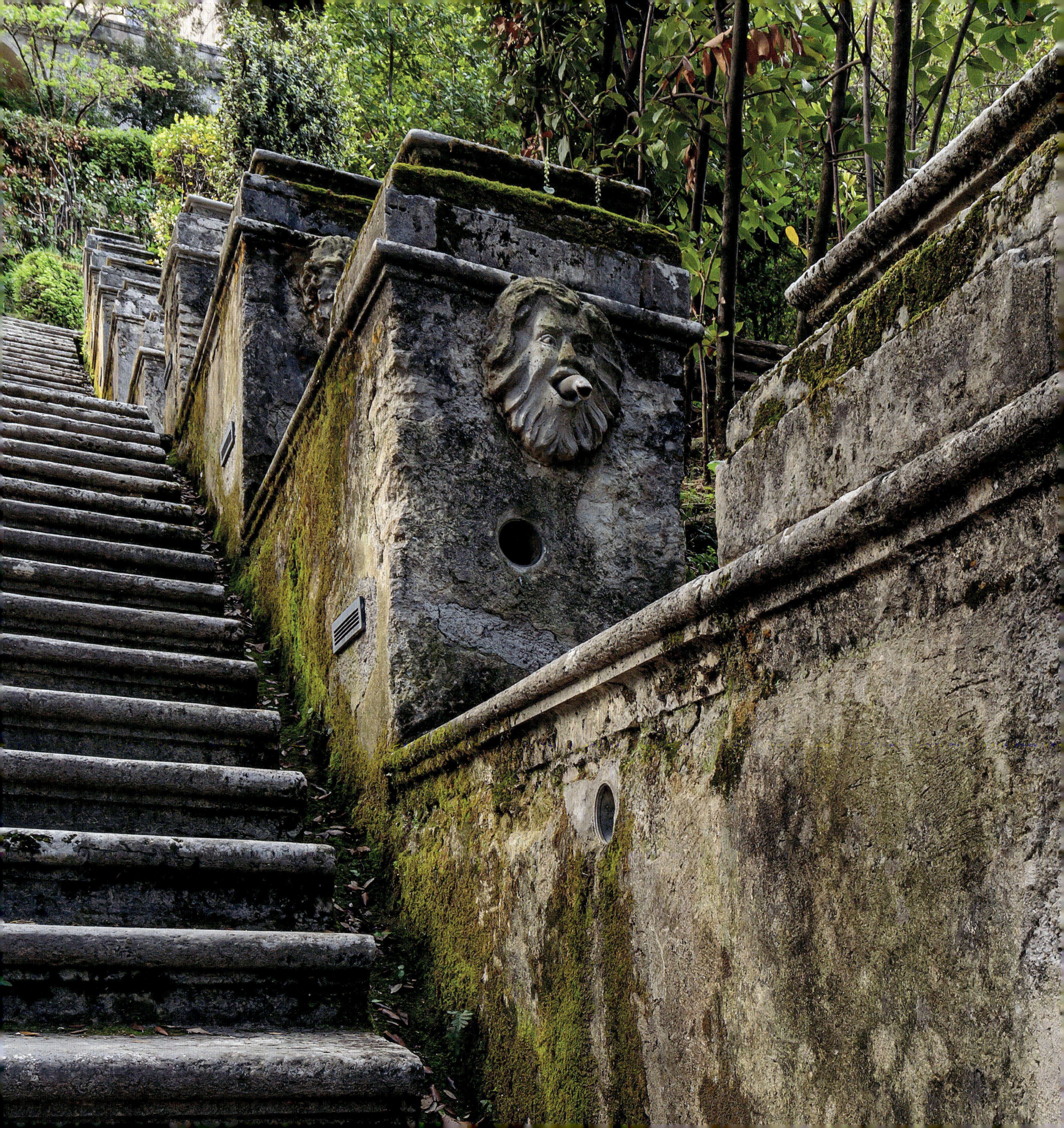

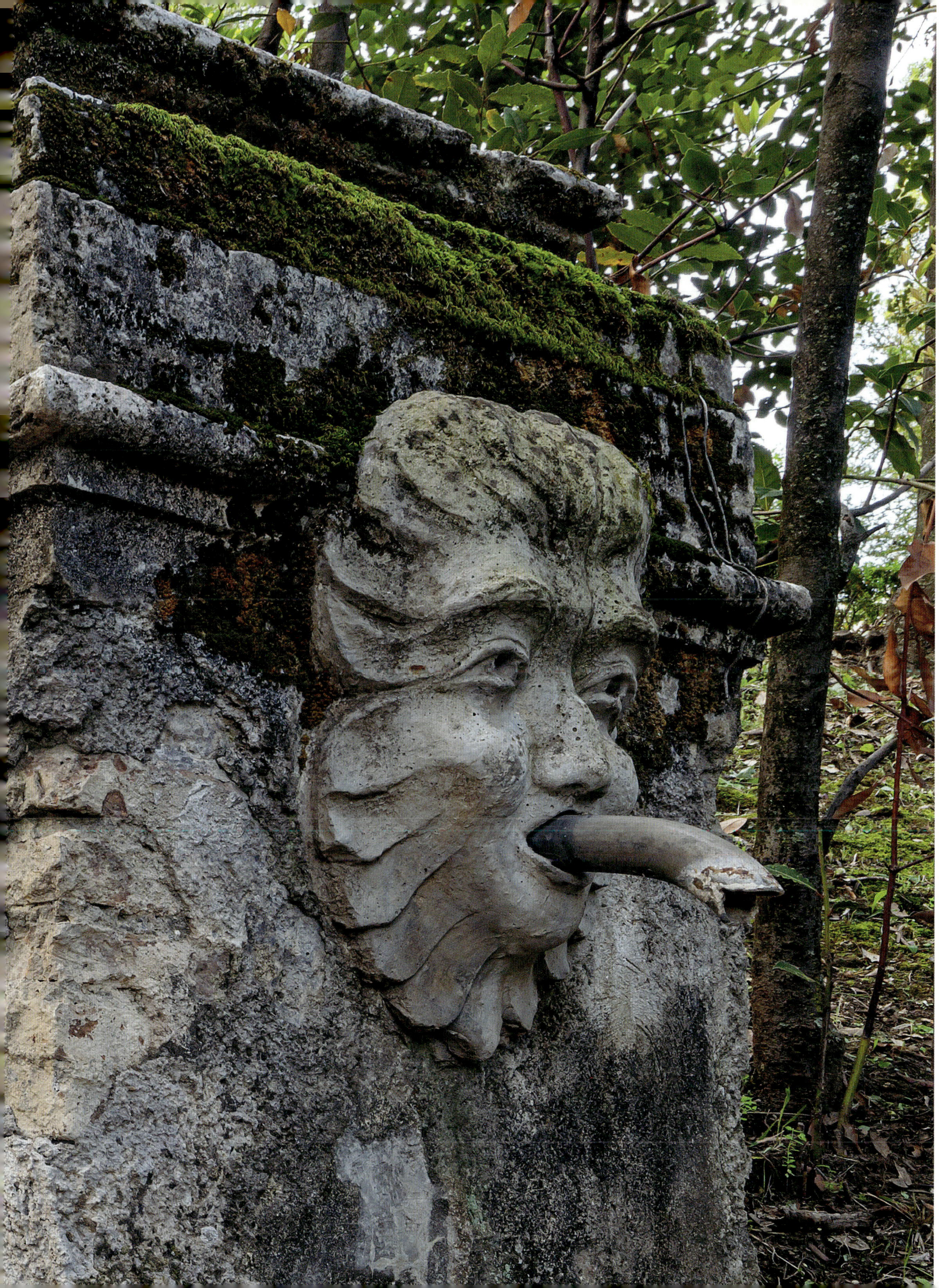

Detail of a mask/spout
of the Bollori Staircase

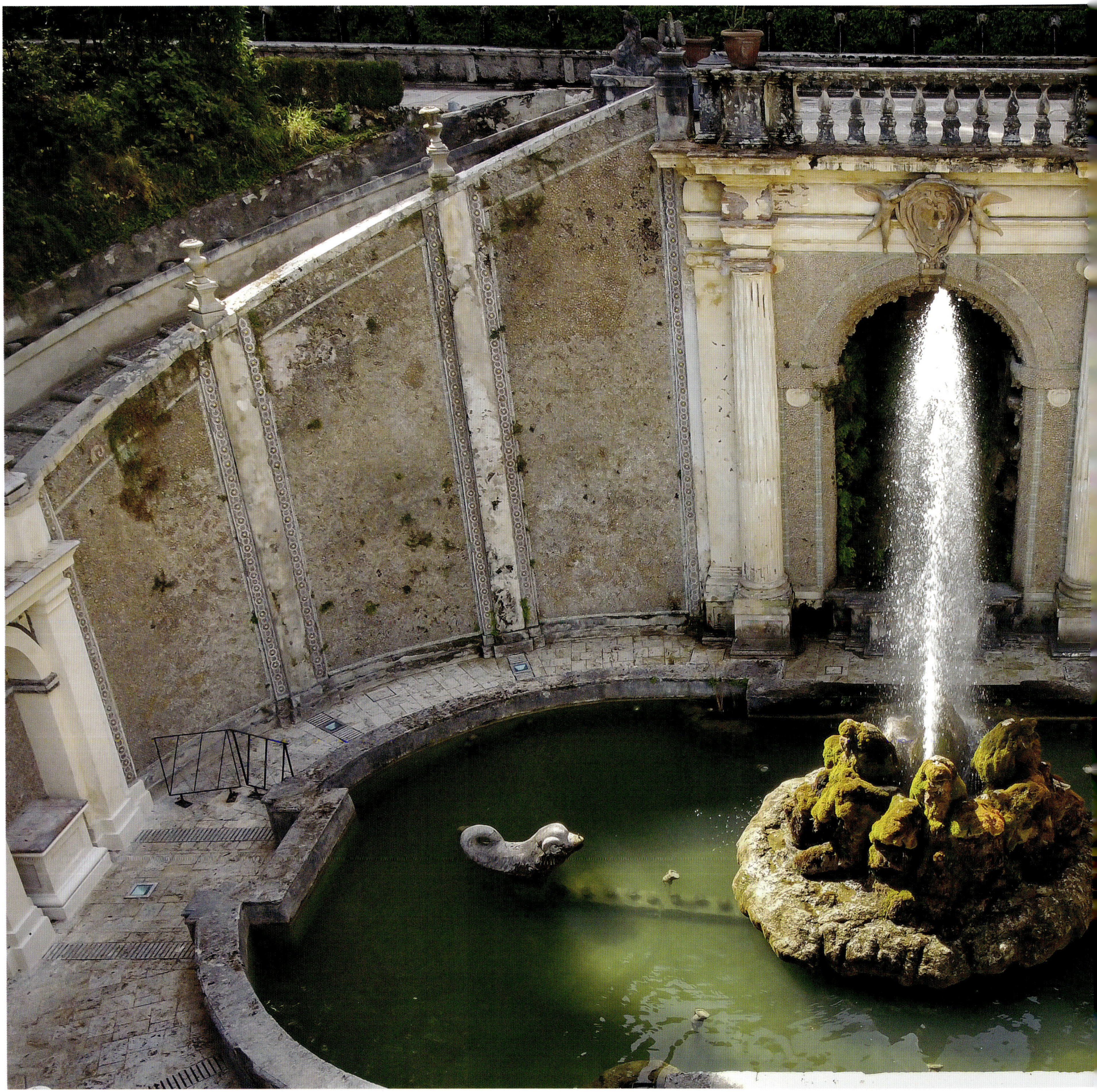

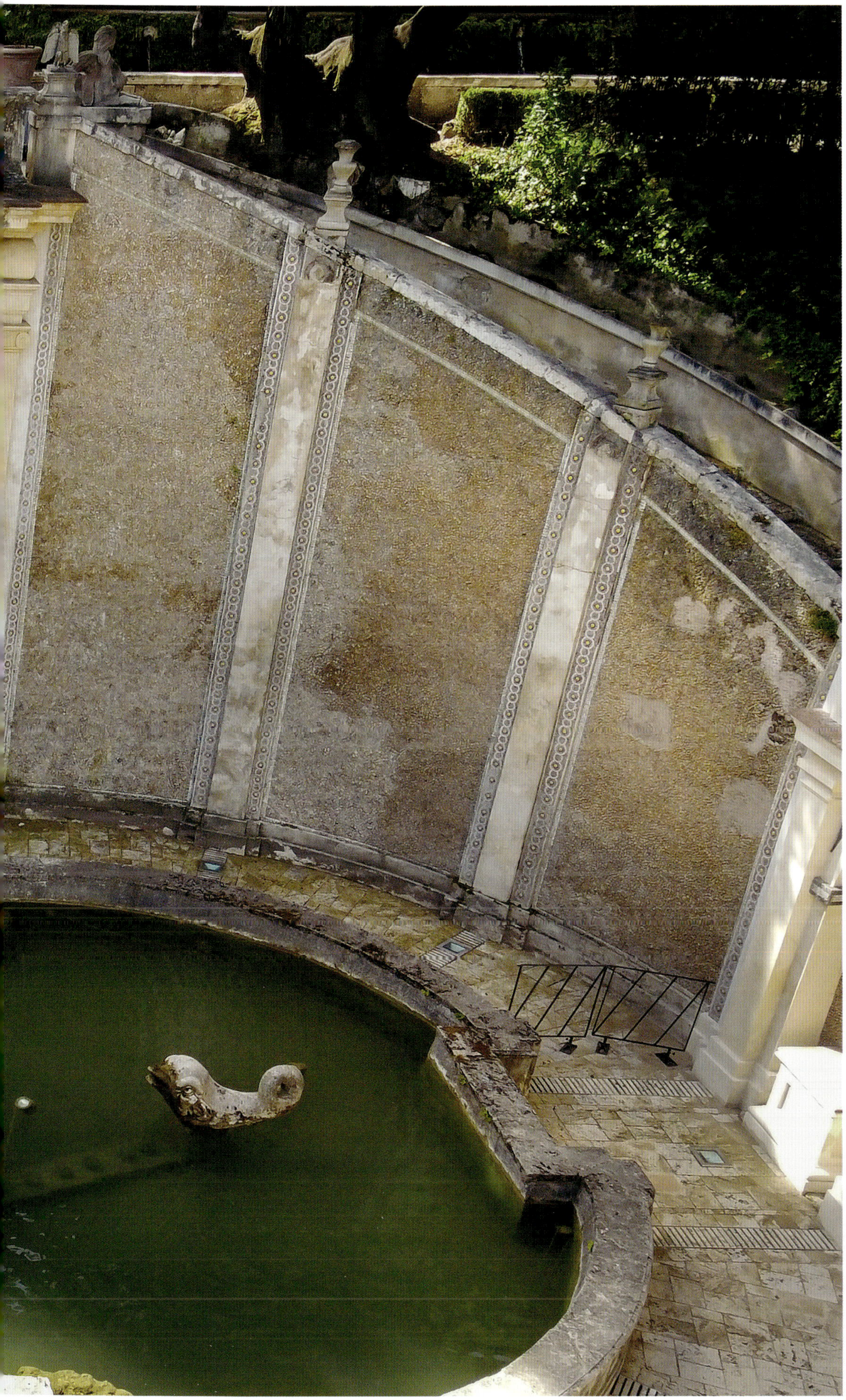

Pirro Ligorio, Fountain of the Dragons, 1572
The fountain was also known as the Fountain of the Pinwheel, because of the very complicated hydraulic mechanisms and artifices devised by Tommaso Ghinucci, which were able to reproduce, in rapid sequence, a whirlwind of rumbles and explosive noises (gunshots, firecrackers, the thunder of cannons, crackling, explosions and the sounds of arquebuses and other weapons).

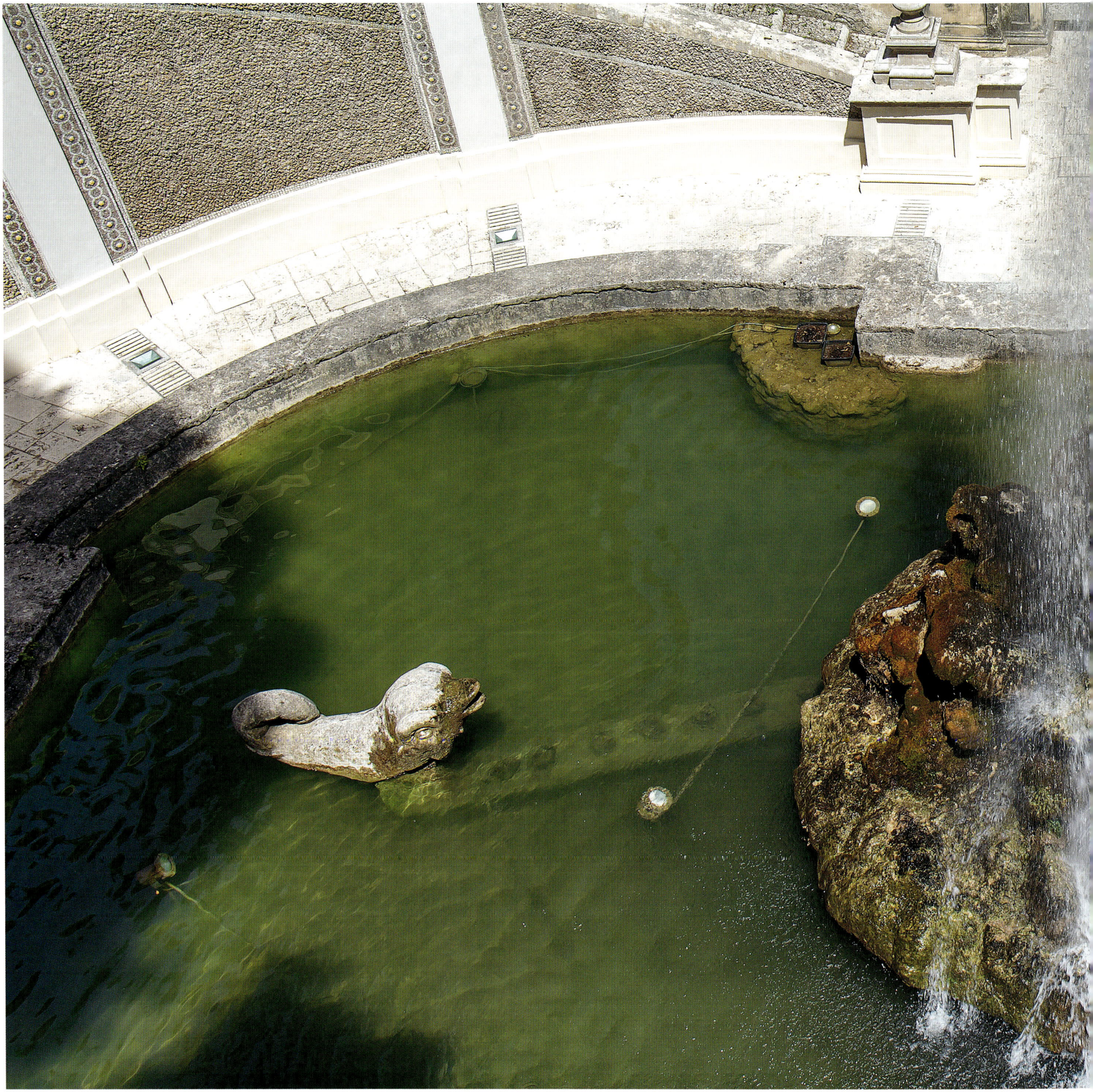

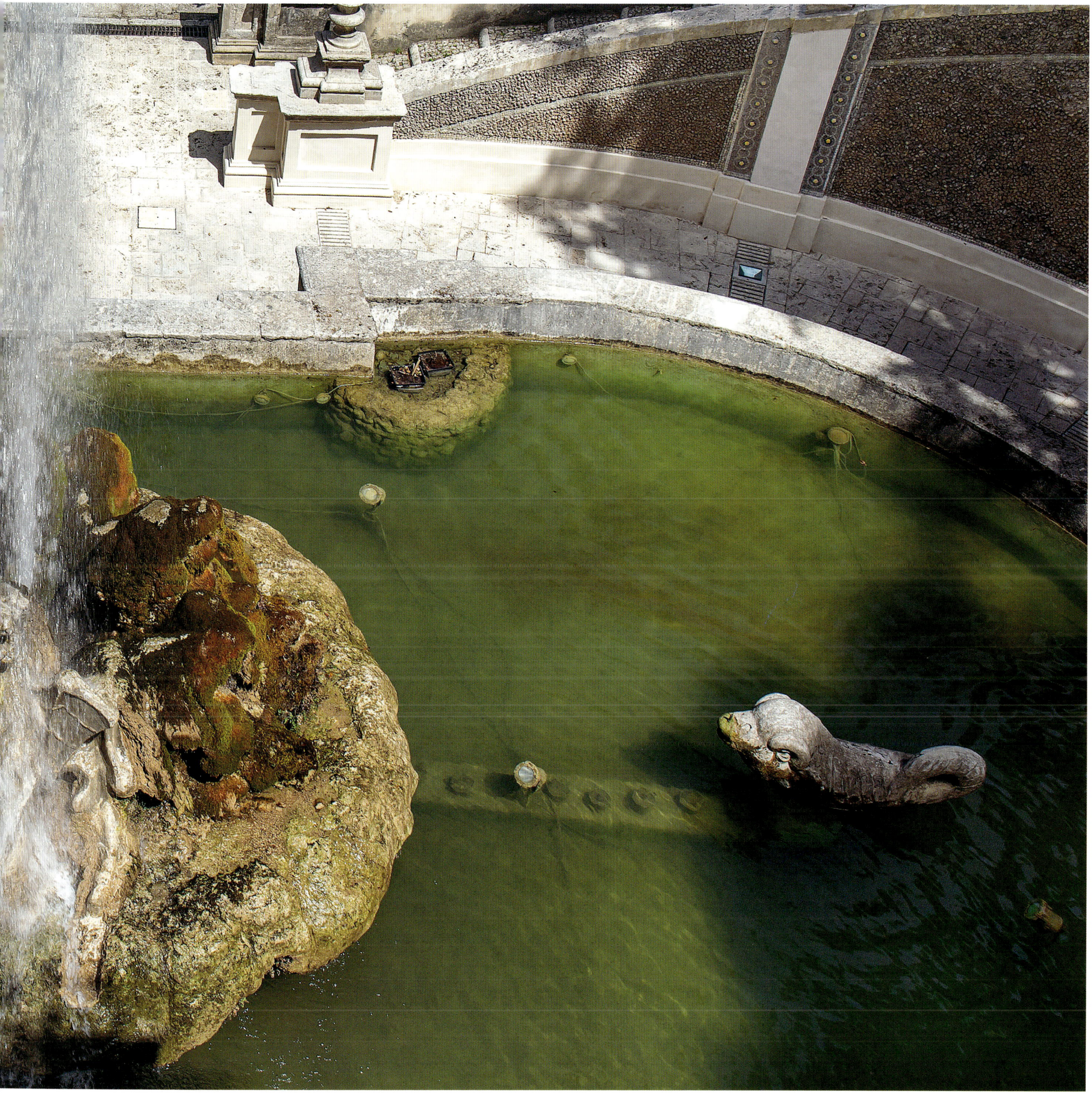

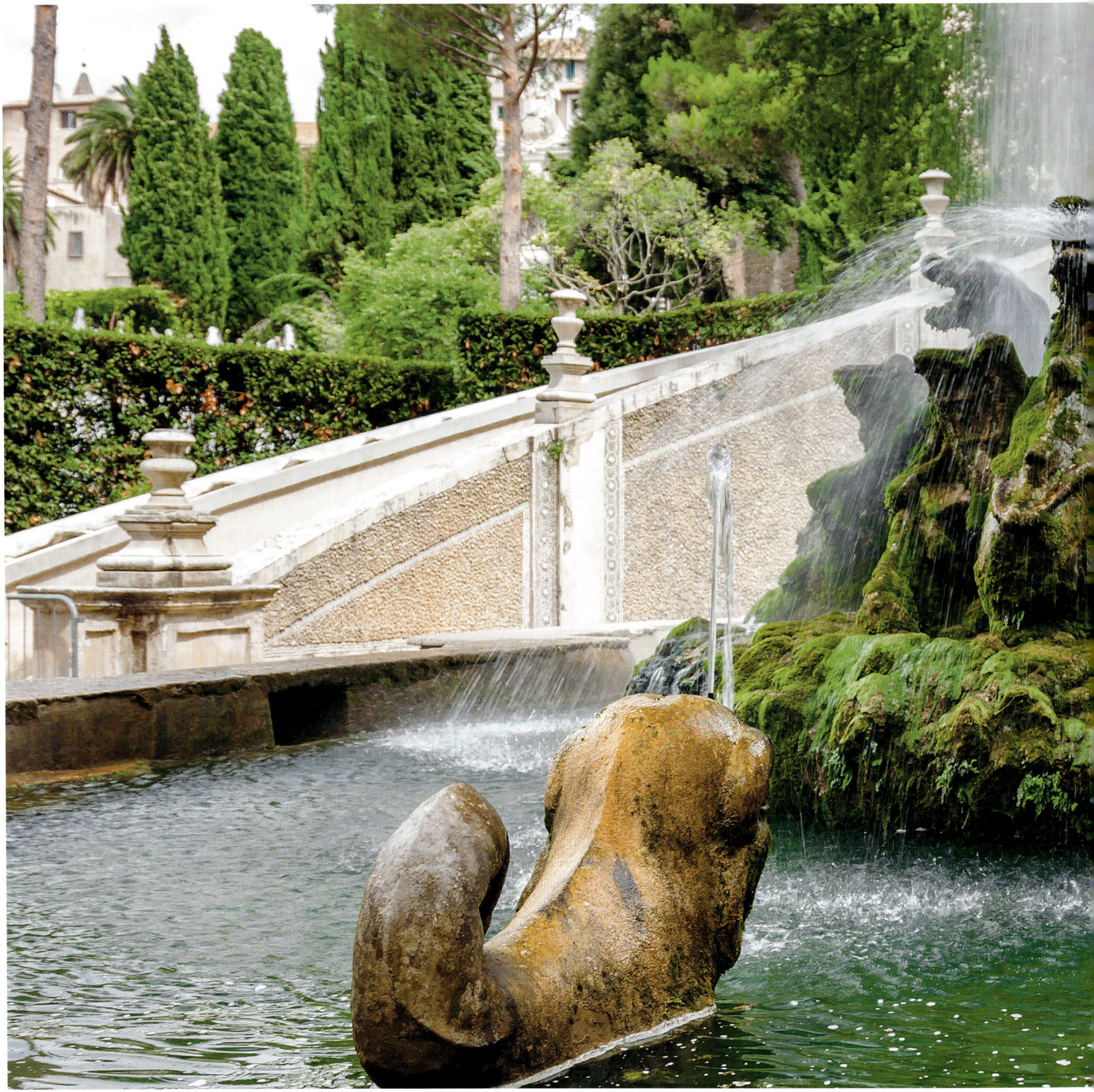

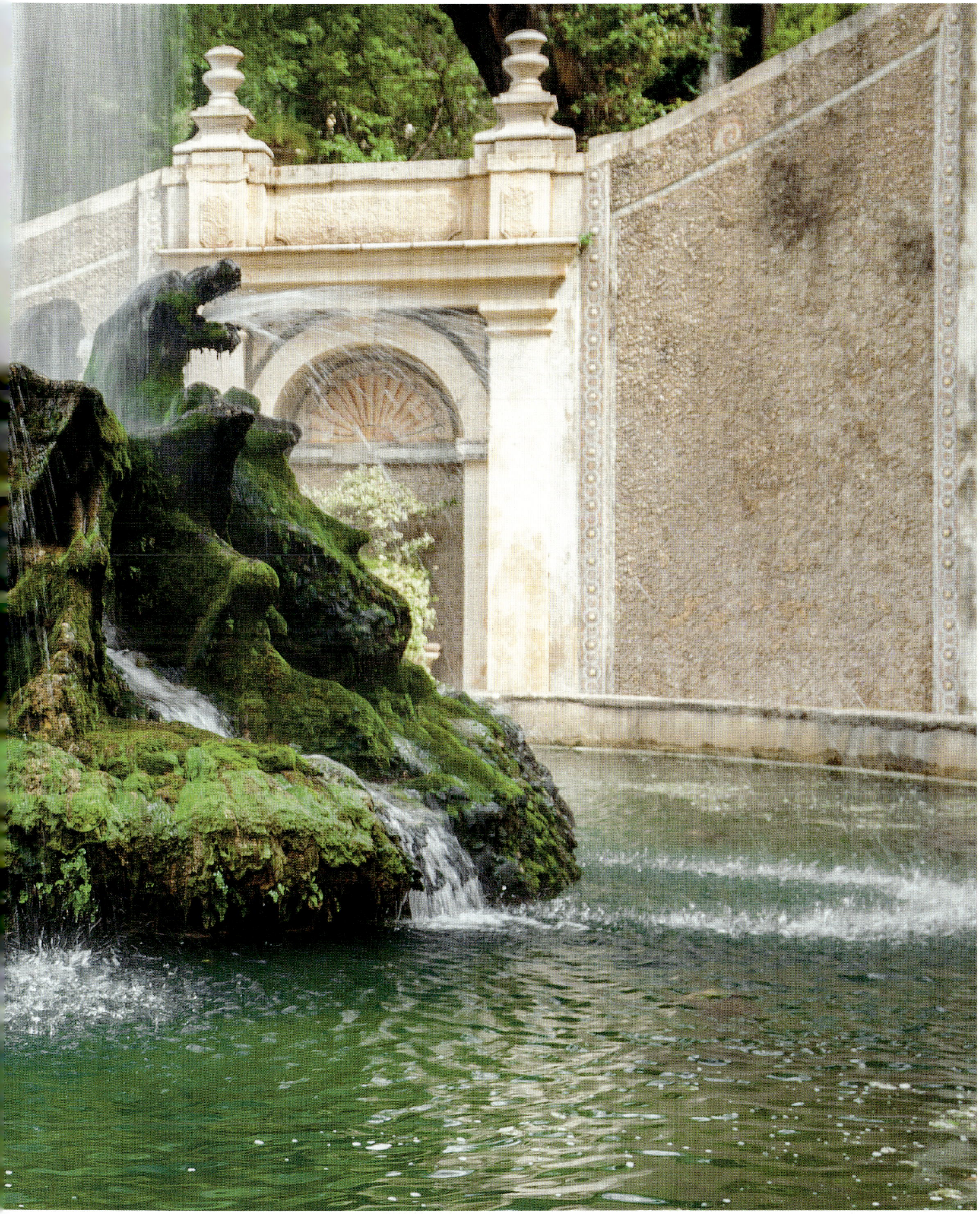

Here and on pages 132–133
Fountain of the Dragons
It is located at right-angles to the palace and the entrance to the valley. In Pirro Ligorio's design, it was one of the garden's marvels due to the aforementioned water and sound artifices, representing a magnificent commentary on the iconographic theme played out in the niches of the masonry surrounding the area. This theme celebrated the enterprise of Hercules, mythical progenitor of the d'Este family, in the Garden of the Hesperides, and thus by association to Ippolito II, exponent of the lineage and founder of the villa.

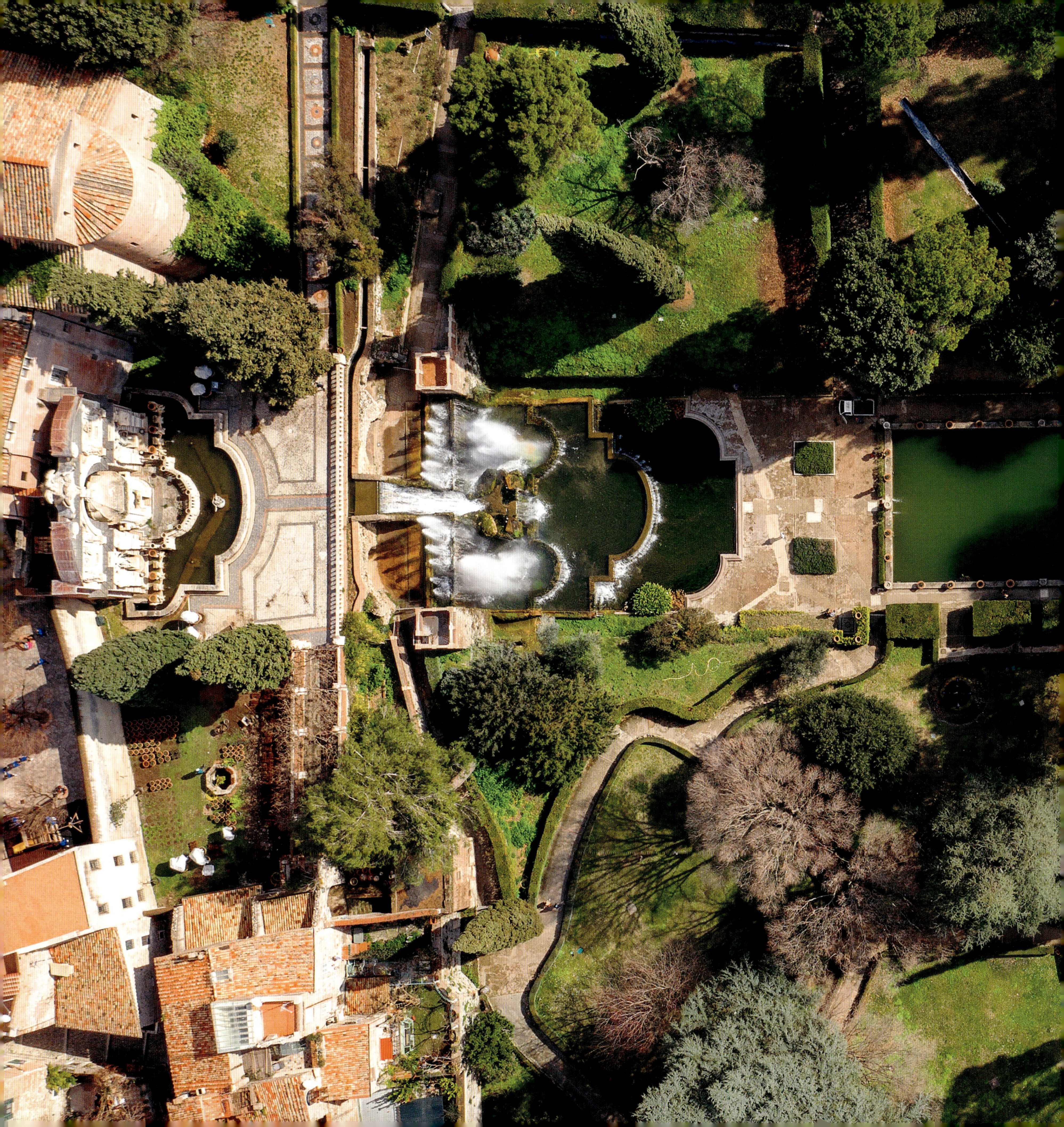

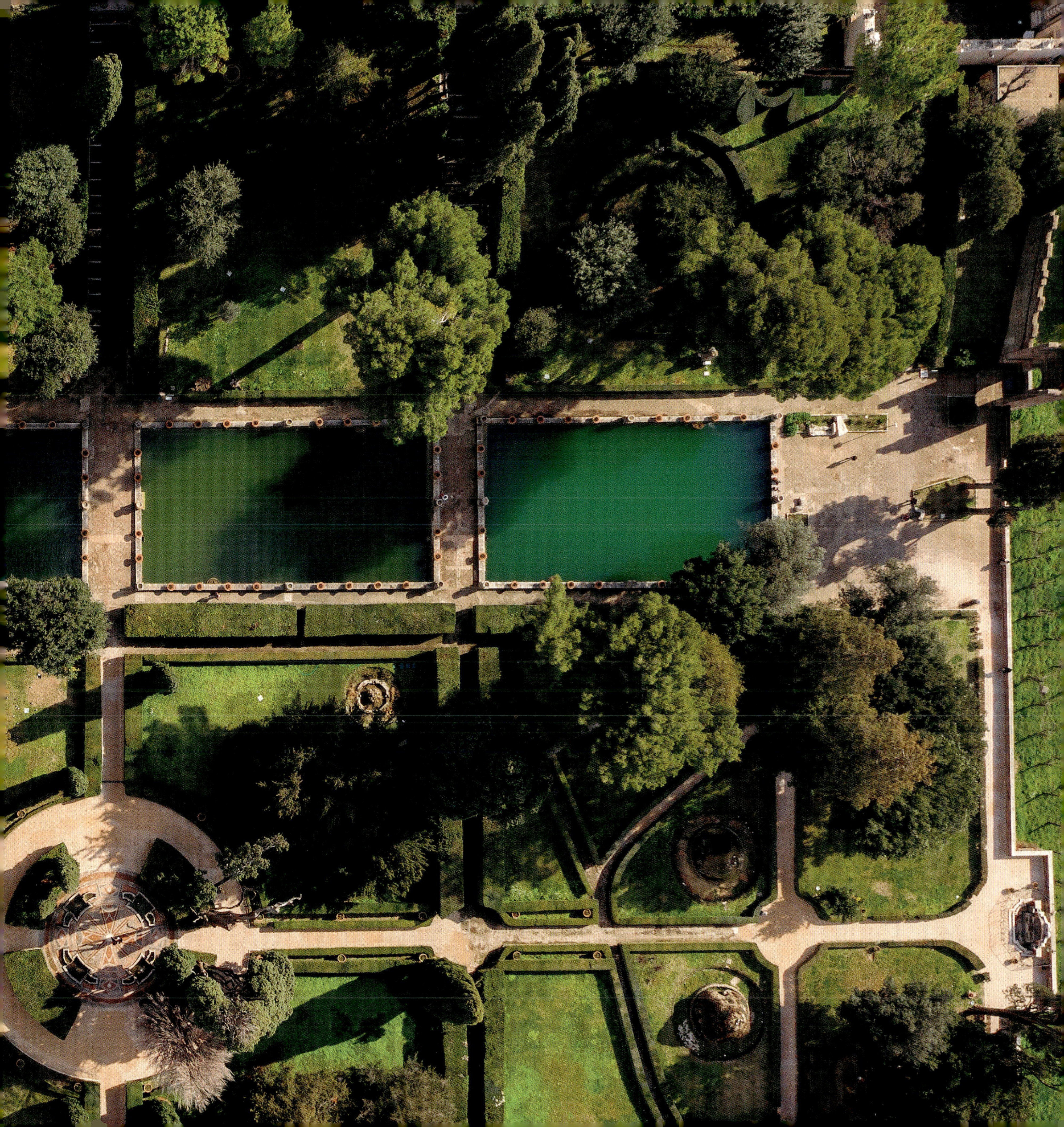

Gregory XIII, since a winged dragon features in the newly elected pontiff's coat of arms. This mythological animal was certainly related to the House of Este through the motto, taken from Ovid's *Metamorphoses*: 'Ab insomni non custodita dracone' (already seen in the Hall of Arts and Crafts), which closely reflects Hercules' eleventh task. Leaving the enclosure of the Fountain of the Dragons and descending the central ramp of the Bollori group, one enters the Peschiere area.

Fish Pools

This vast area, also known as the Flat Garden, accommodates three large rectangular pools, each deep and bordered by kerbs, featuring eight spouts. Originally, as can be seen from Duperac's view, there were four basins and from the two central ones rose two features that, seen from the entrance downstream, formed the lateral limits to the visual framing of the palace. The symmetrical design attributed to this area, now less discernible, most probably focused on the eastern part. Historical evidence attests to the presence of substantial substructures built to shape the slope with three main terraces, each adorned with sumptuous fountains. However, only the two fountains on the upper terraces were constructed, while the one at the bottom, later known as the Neptune Fountain and erected in the seventeenth century, remained incomplete.

Neptune Fountain

The fountain features a monumental hydraulic apparatus with high jets arranged on several levels and symmetrically divided by a scenic waterfall. It stands on top of sixteenth-century substructures that were remodelled in the seventeenth century with apparatuses of faux 'sweating' rocks intersected by the waterfall that pours water over a grotto arranged in the middle of a basin. Many have assumed that the seventeenth-century arrangement depended on an intervention by Bernini, but this cannot be proven in view of the modest

View of the Fish Pools, in the background are visible the Neptune Fountain and the Fountain of the Organ
The three large basins, which collect the water of the Neptune Fountain, were originally a grandiose waterway that transported the waters coming from the Fountain of the Organ above towards the Fountain of the Sea, which was never realised.

Pages 136–137
Aerial view of the garden

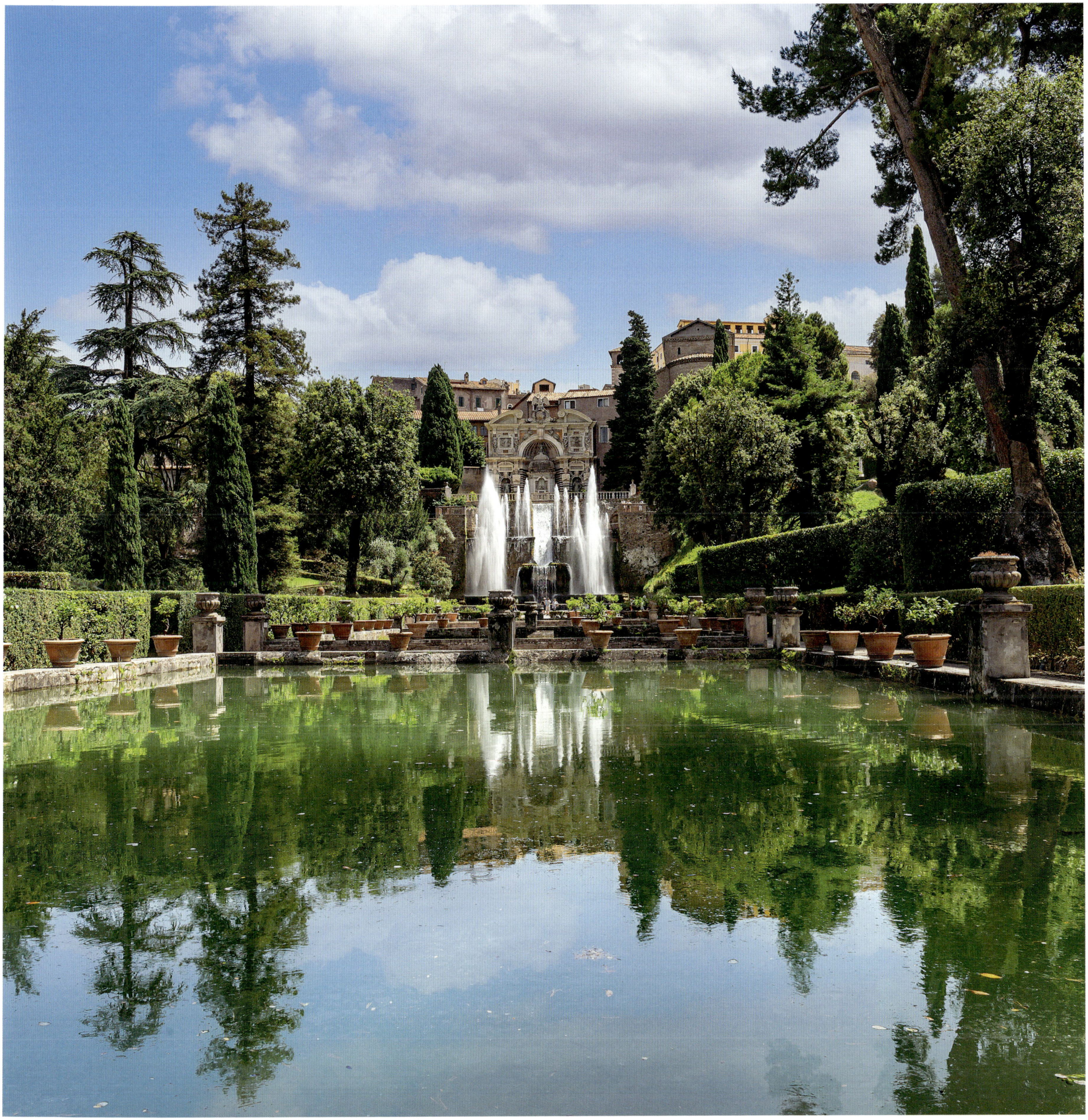

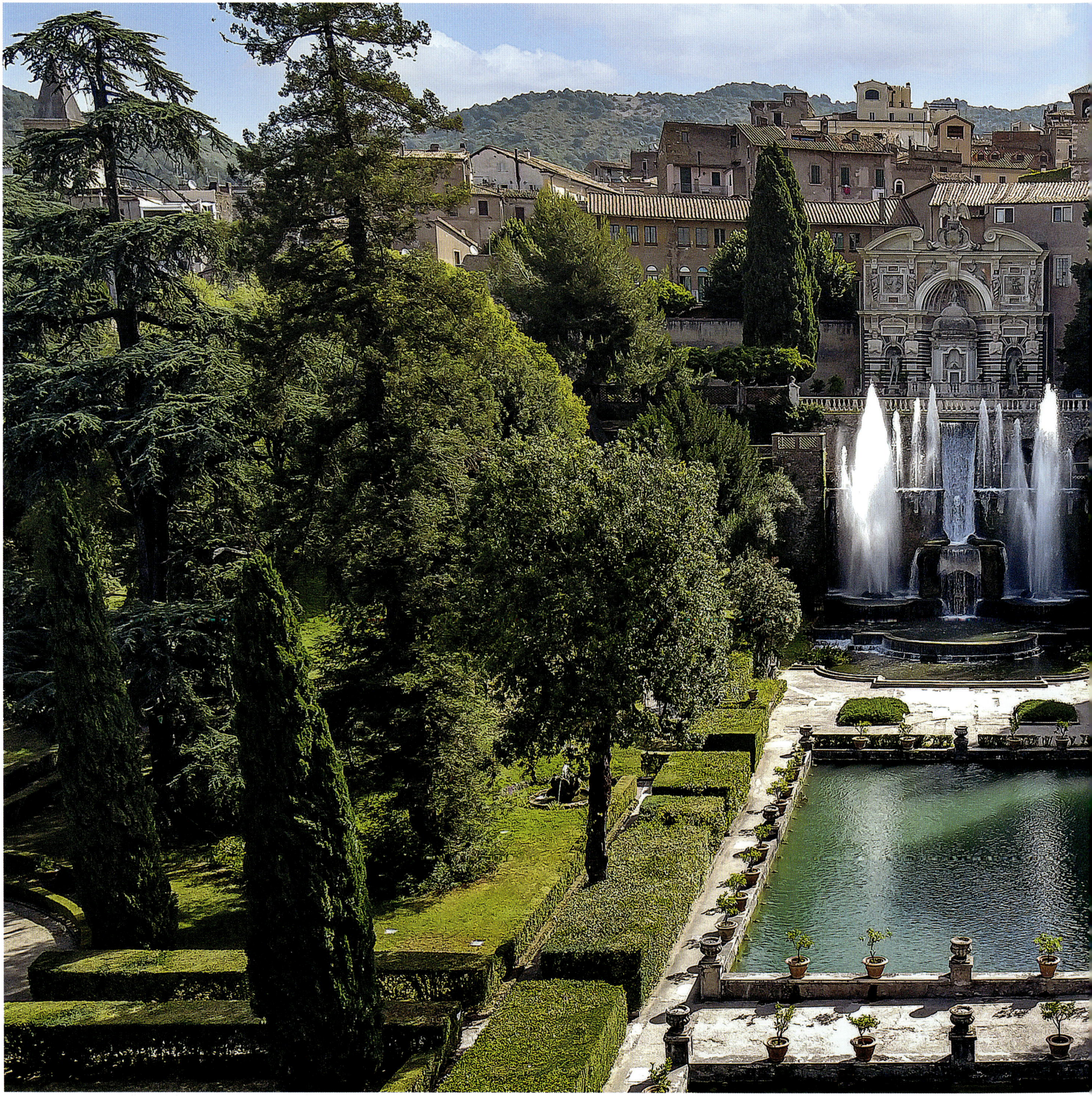

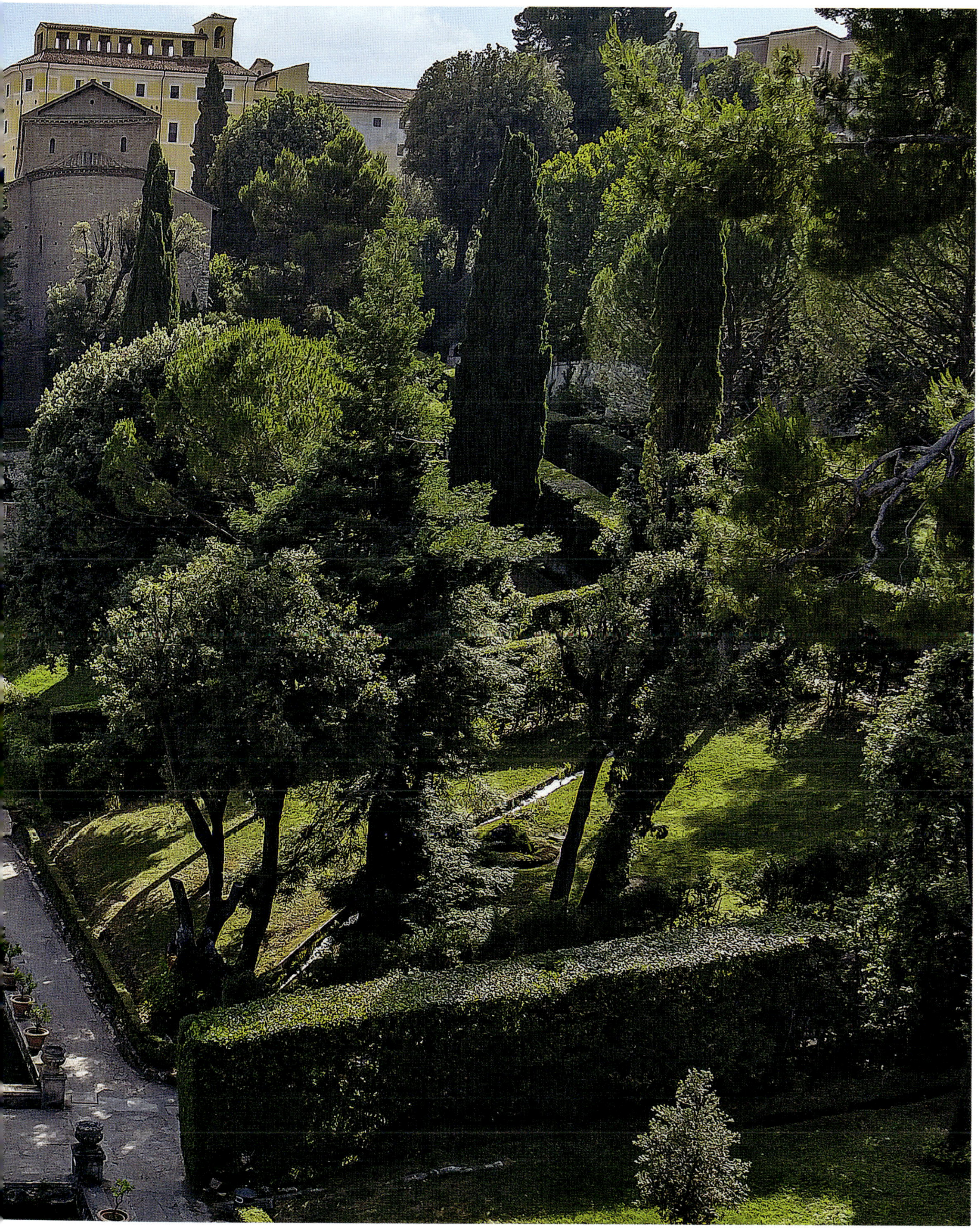

View of the Peschiere, or Fish Pools,
with the Neptune Fountain and Fountain
of the Organ in the background

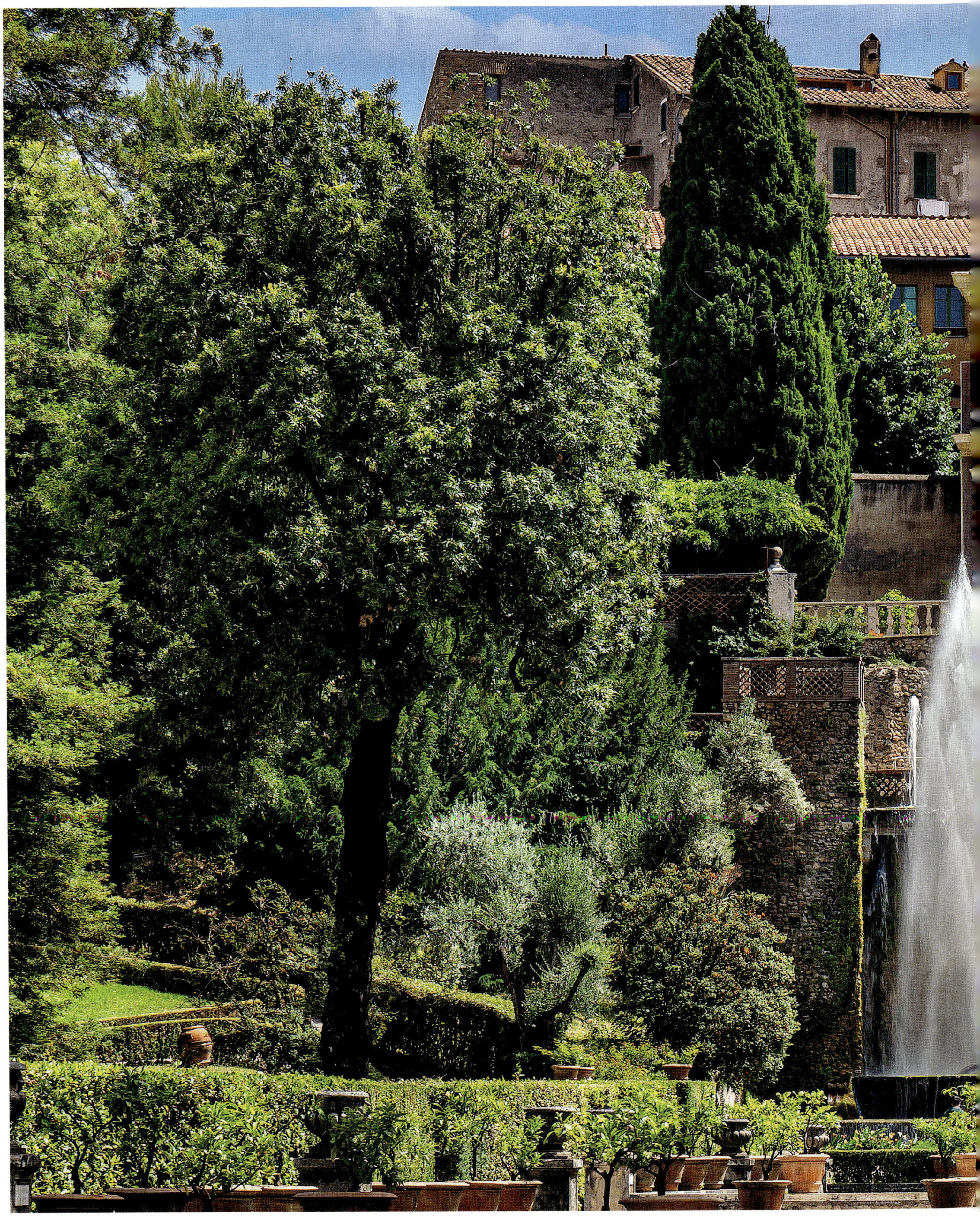

The Neptune Fountain with the Fountain of the Organ in the background

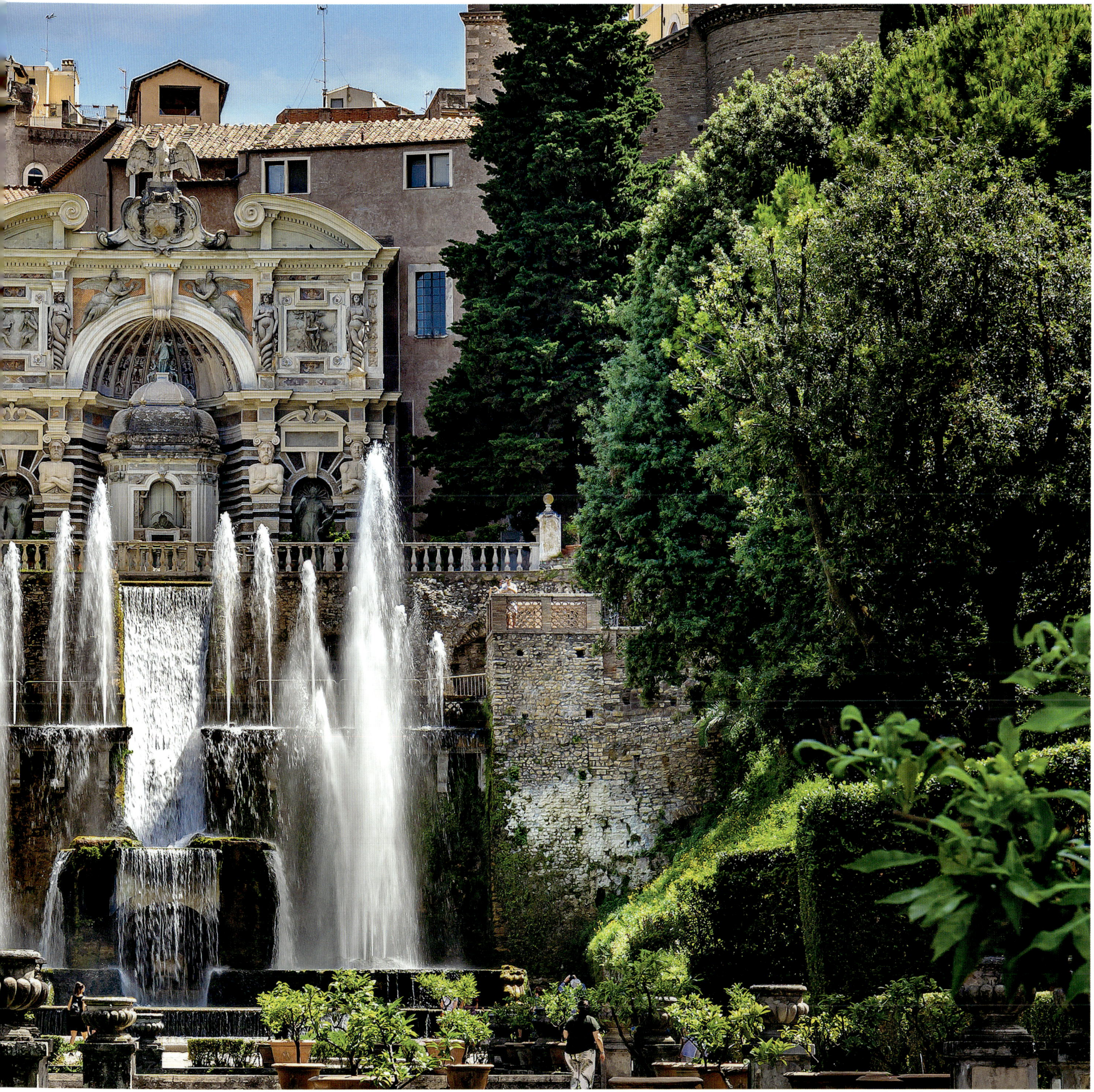

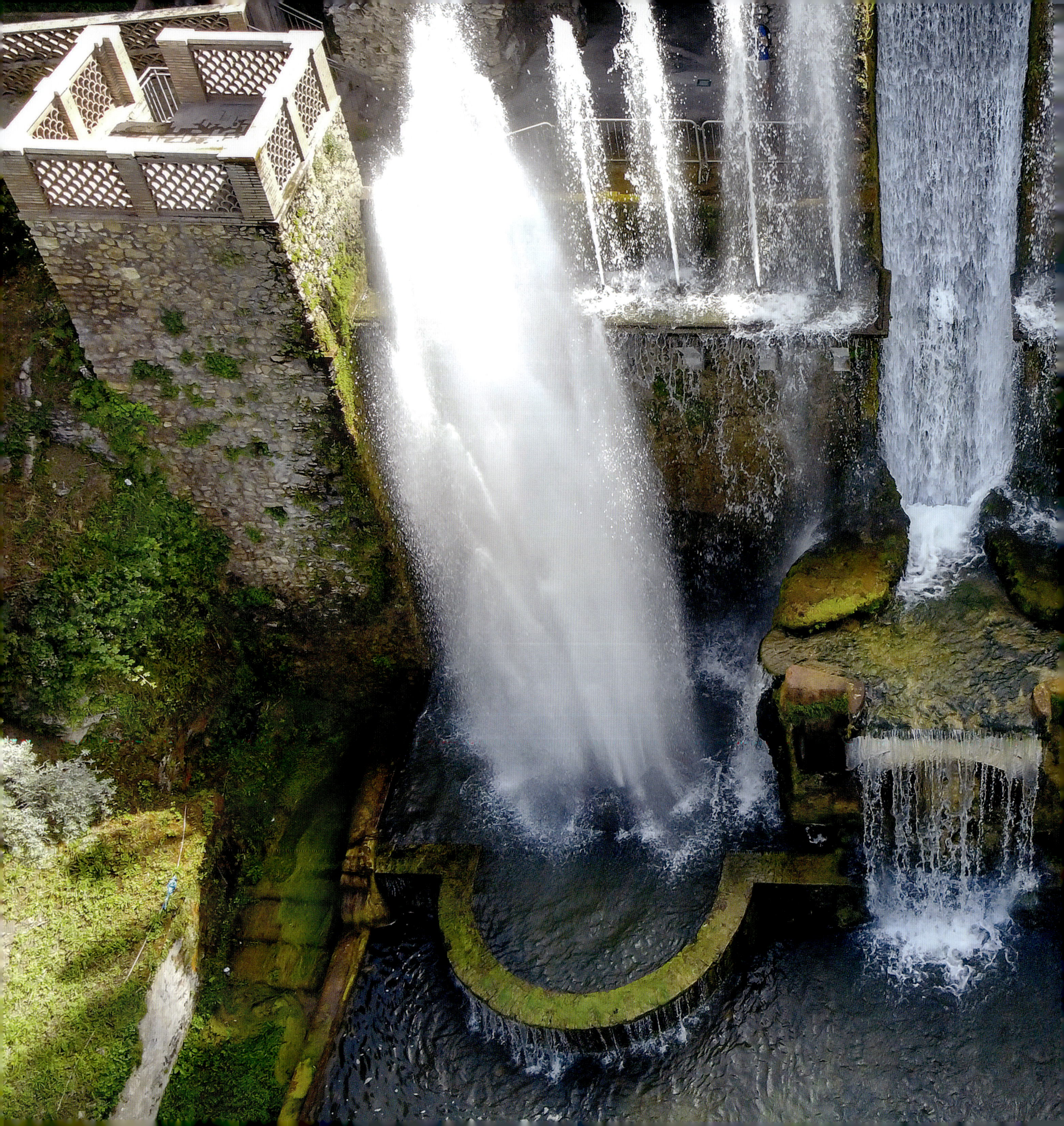

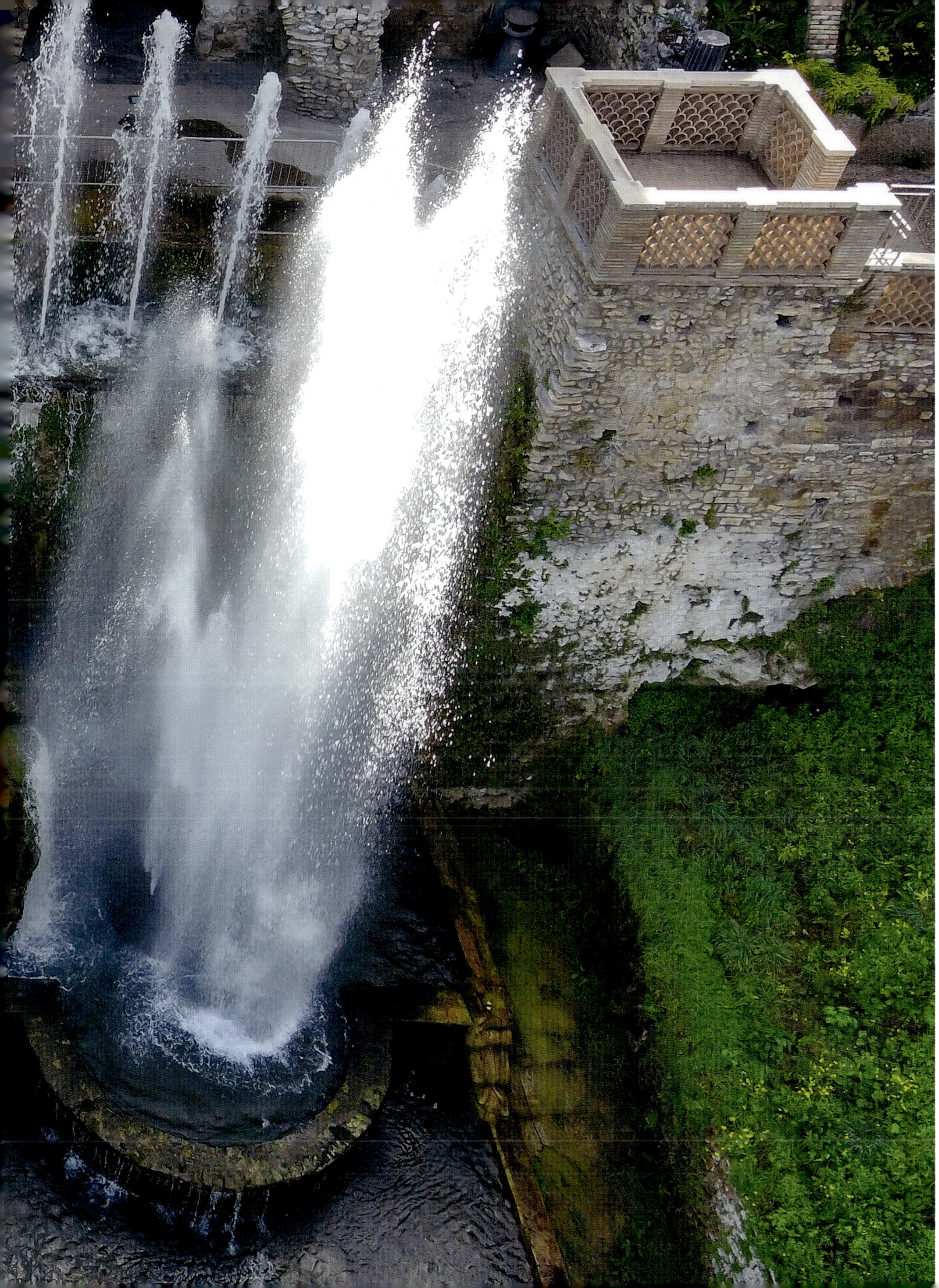

Attilio Rossi, Emo Salvati,
Neptune Fountain, 1927

Pages 146–147
View of the Fish Pools

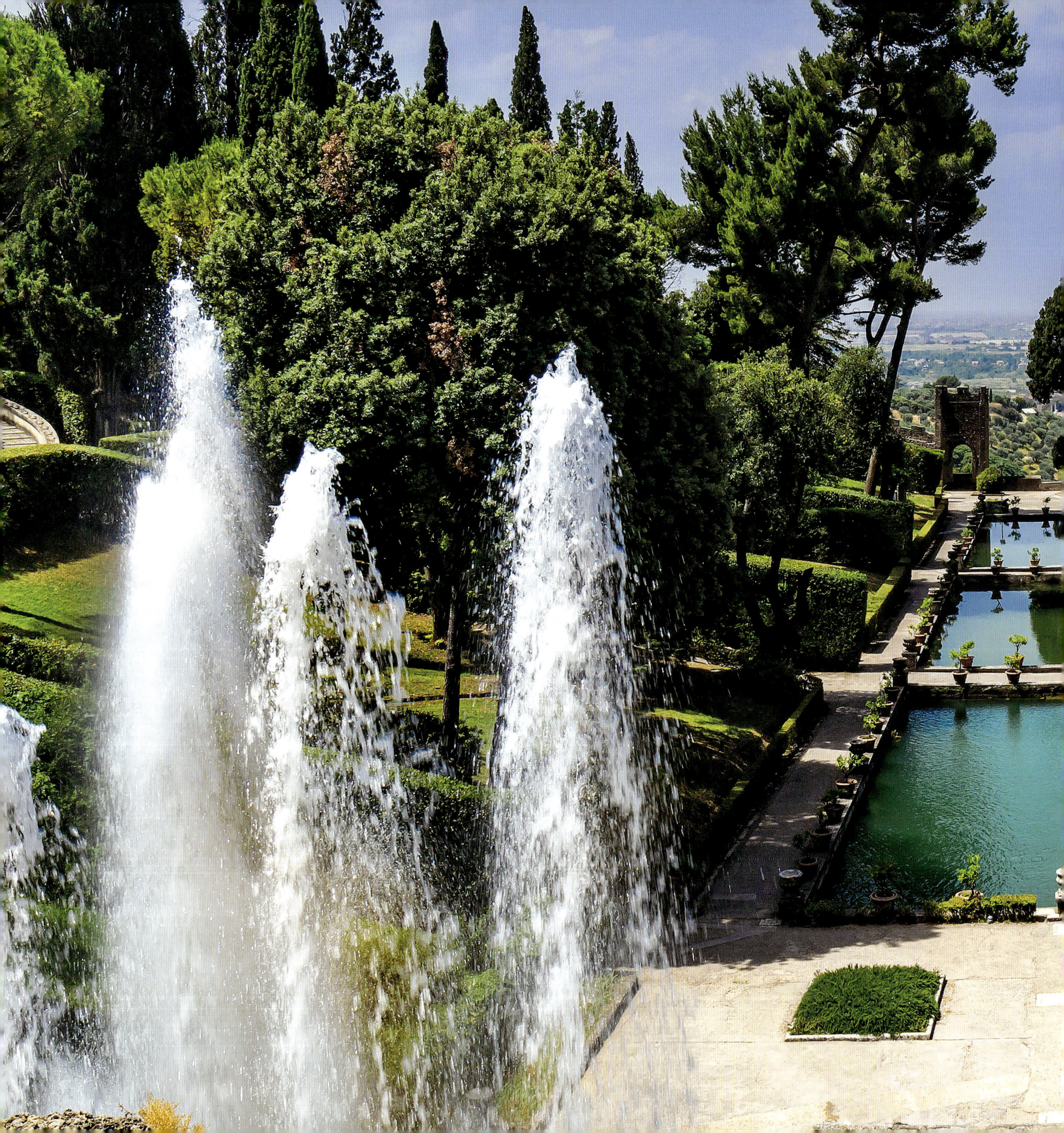

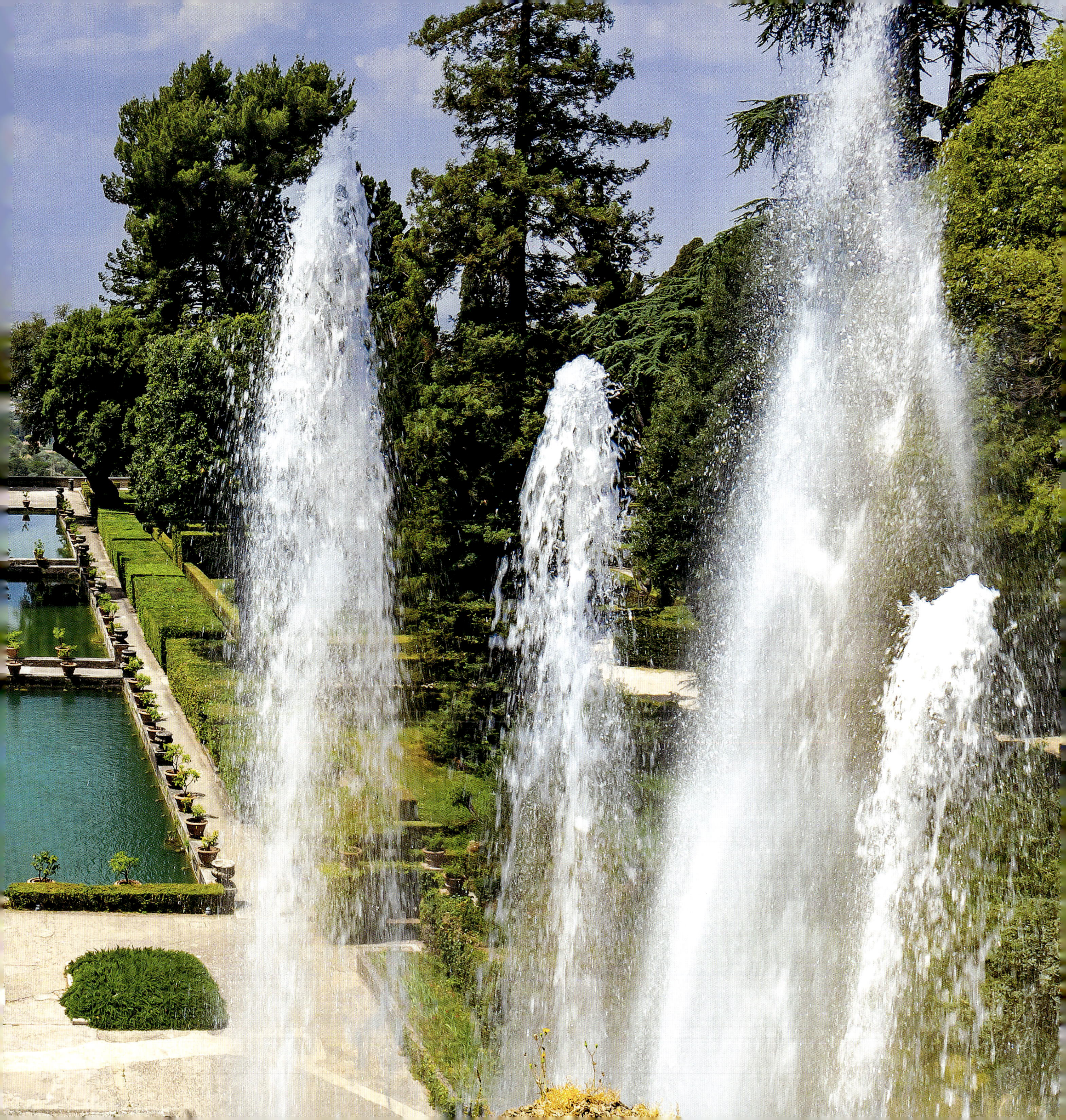

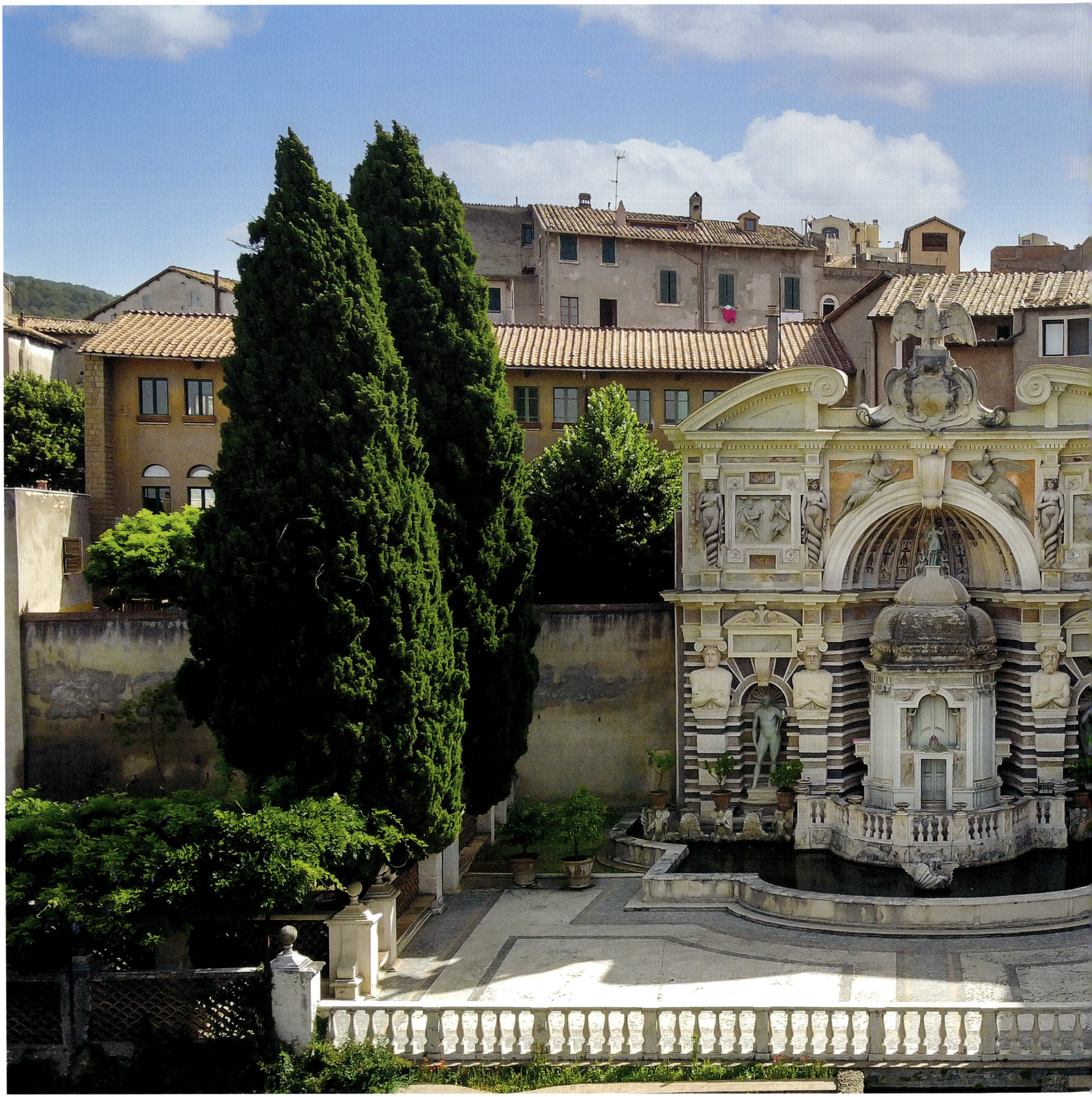

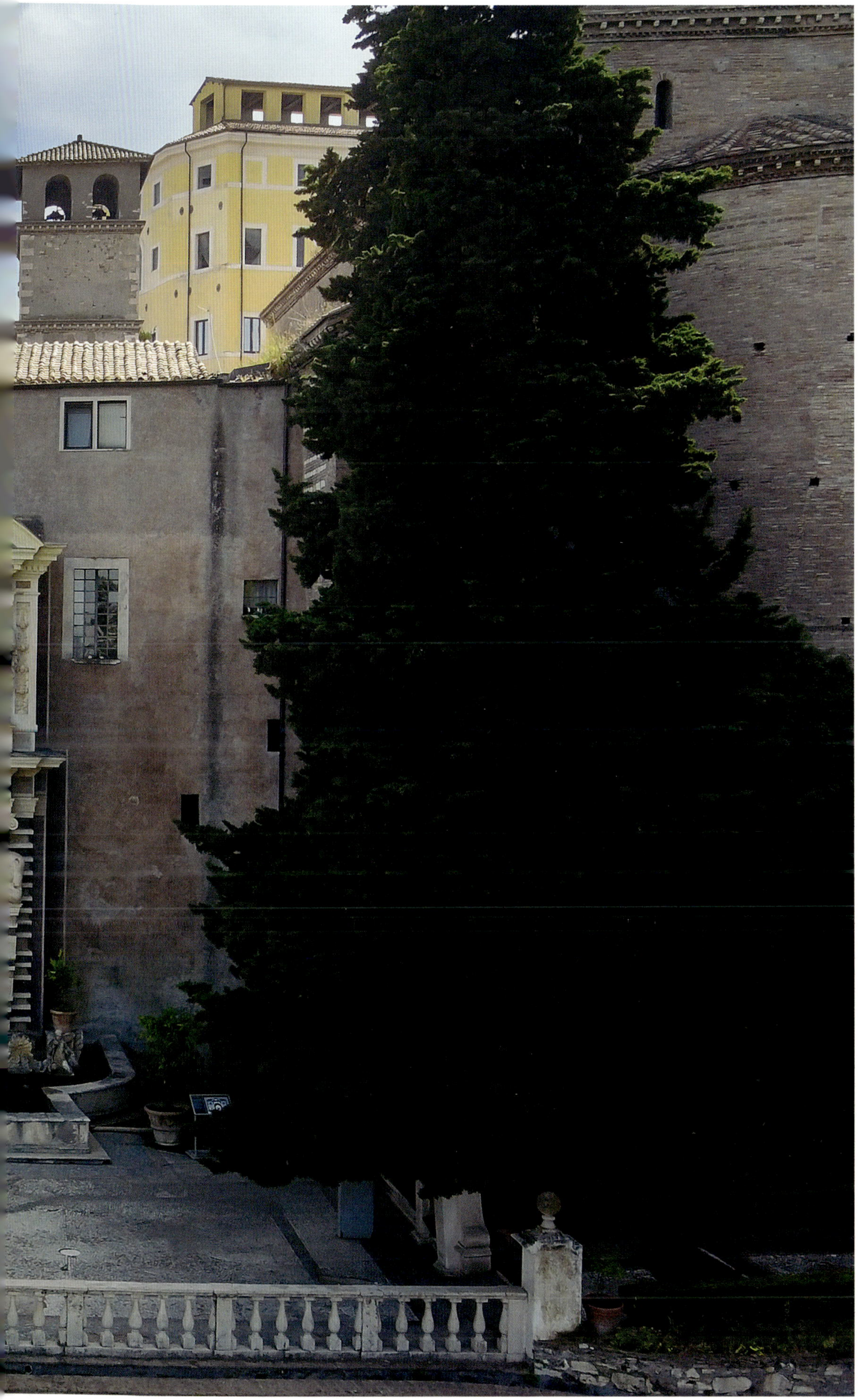

Fountain of the Organ, 1566–early decades of the seventeenth century, designed by Luc Leclerc and Claude Venard
The current name of the fountain, originally called del Diluvio (of the Flood) or della Madre Natura (of Mother Nature) for the statue later moved along the garden wall, results from the presence of an organ and hydraulic devices already in operation in 1572 for the visit of Pope Gregory XIII. It soon became famous and was immediately imitated in the villas of Pratolino and Quirinale. The monumental façade actually constituted the scaenae frons *of a spectacle produced by water and the sounds originated by the complex hydraulic mechanisms.*

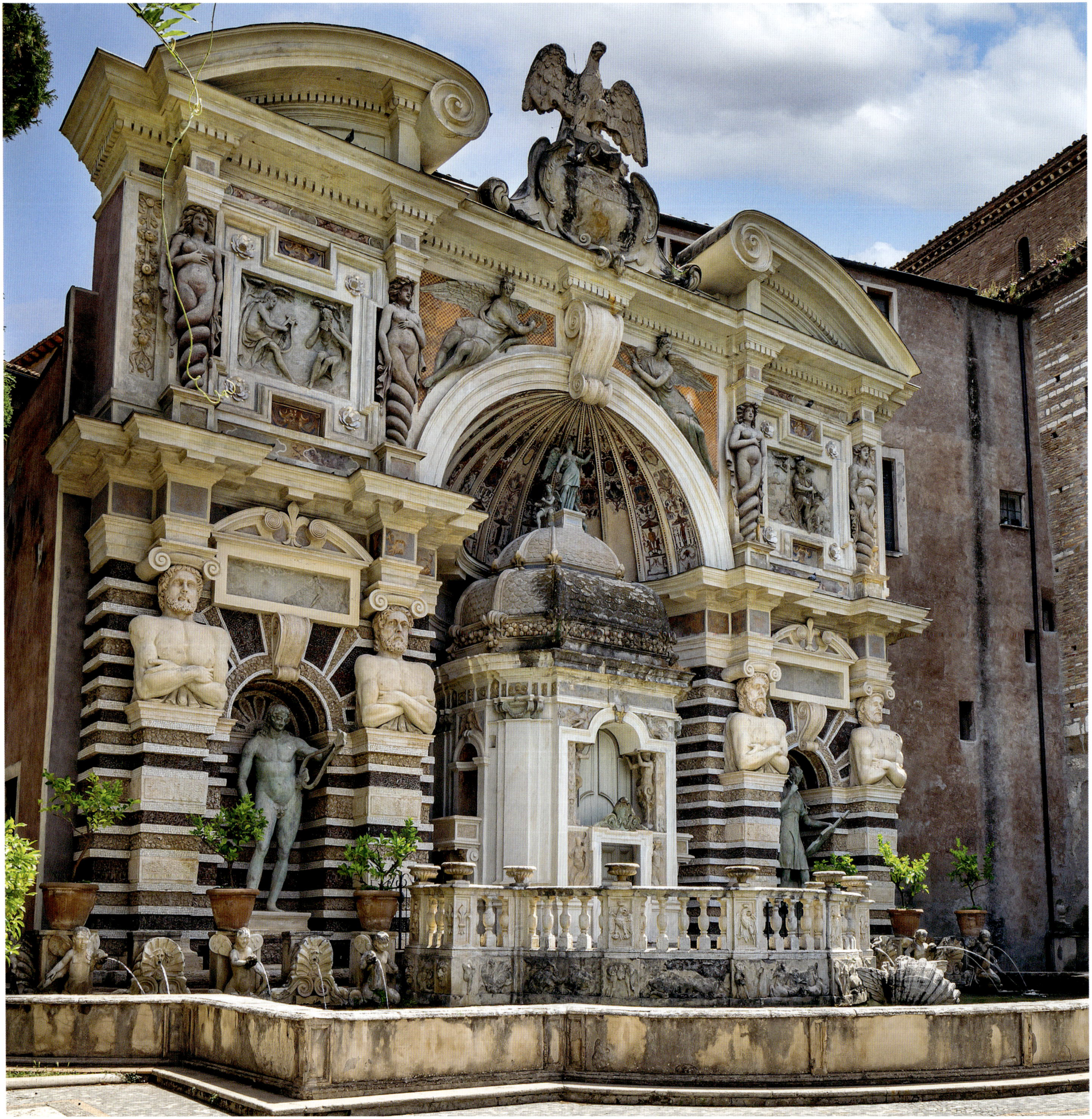

layout shown in Venturini's view. In the second two decades of the twentieth century, commissioned by Attilio Rossi, engineer Emo Salvati realised the fountain's current form, using reinforced concrete to give life to the project. This technique, which was still innovative at the time, allowed him to create large basins, projecting on imposing modillions (projecting brackets under the top of a cornice) in the upper part of the fountain, within which the water gushes upwards by means of a hydraulic system of metal pipes, and then overflows, giving the visual effect. Inside the basin, the jets reach imposing heights, making the ensemble particularly impressive, so much so that they are among the most photographed in the entire garden. At a slightly higher level than the basins, we come to the Grottoes of the Sibyls.

Grottoes of the Sibyls

This consists of three cave nymphaea, arranged in succession and built within the substructures. In the middle part, the passage is surmounted by the waterfall of the Neptune Fountain, which derives its water from the upper level, where the Fountain of the Organ is located.

Fountain of the Organ

Towering over the Neptune Fountain, of which it forms the imposing scenic backdrop, the Fountain of the Organ completes the trio of sound fountains commissioned by Ippolito II and executed by French technicians (Luc Leclerc with the help of his nephew, Claude Venard). The name derives from the hydraulic organ, an instrument invented by Heron and later imported to Rome and Byzantium. The mechanism realised at Villa d'Este, unlike the ancient one (which functioned by means of a container filled with air into which water was introduced, thus compressing the air, causing it to escape and forming sounds) was based on a flow of water that reached the terrace of the fountain and was then introduced

Fountain of the Organ, 1566–early decades of the seventeenth century, designed by Luc Leclerc and Claude Venard
The fountain was unfortunately subject to rapid deterioration and as early as the beginning of the seventeenth century Cardinal Alessandro d'Este had it restored. On this occasion, he enriched the decorative apparatus, which until then had been limited to four travertine telamons. The stucco statues of Apollo and Orpheus were placed in the niches of the first order, the mermaid-like Caryatids in the second, winged Victories in the pendentives, and bas-reliefs in stucco with mythological scenes related to music at the keystone of the arch. The rounded basin was also realised anew, while inside the niche, the statue of Mother Nature was replaced by the octagonal aedicula to house the organ.

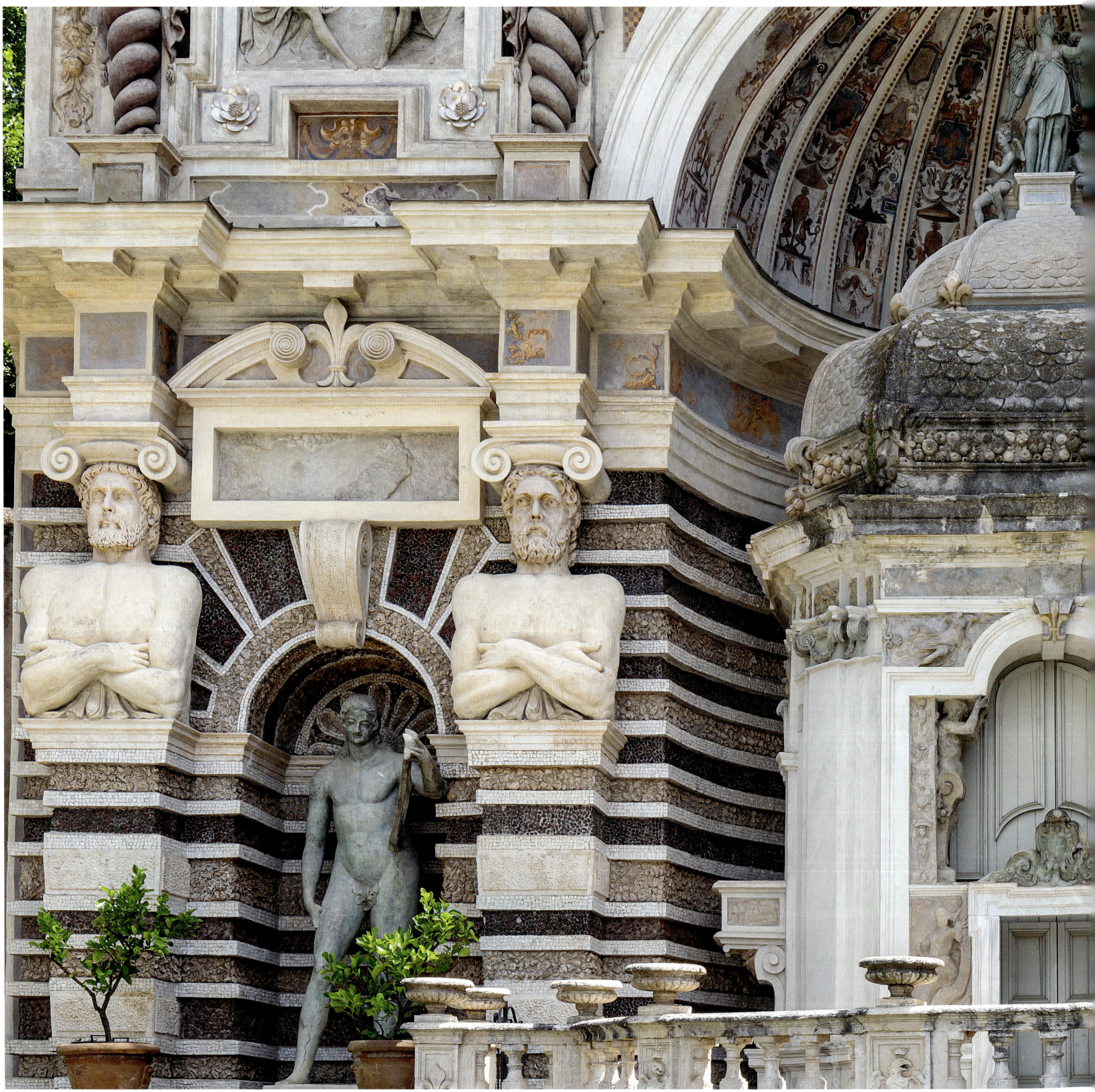

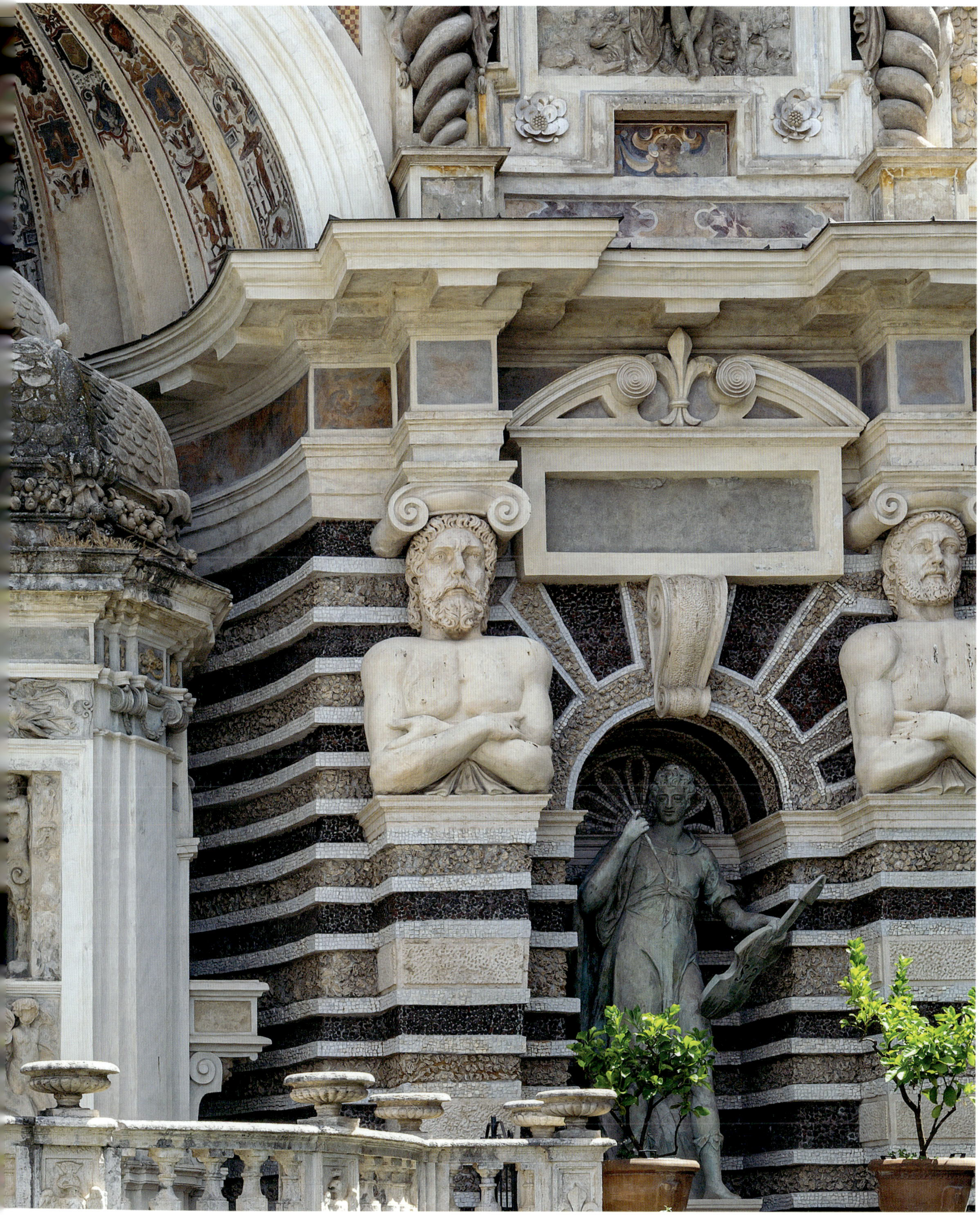

Fountain of the Organ, 1566–early decades of the seventeenth century, designed by Luc Leclerc and Claude Venard
The four colossal telamons are the work by Pirrin del Gagliardo. In the central niche is the small temple that protects the hydraulic organ, designed by Luc Leclerc and Claude Venard.

Pages 154–155
Detail of the top of the Fountain of the Organ

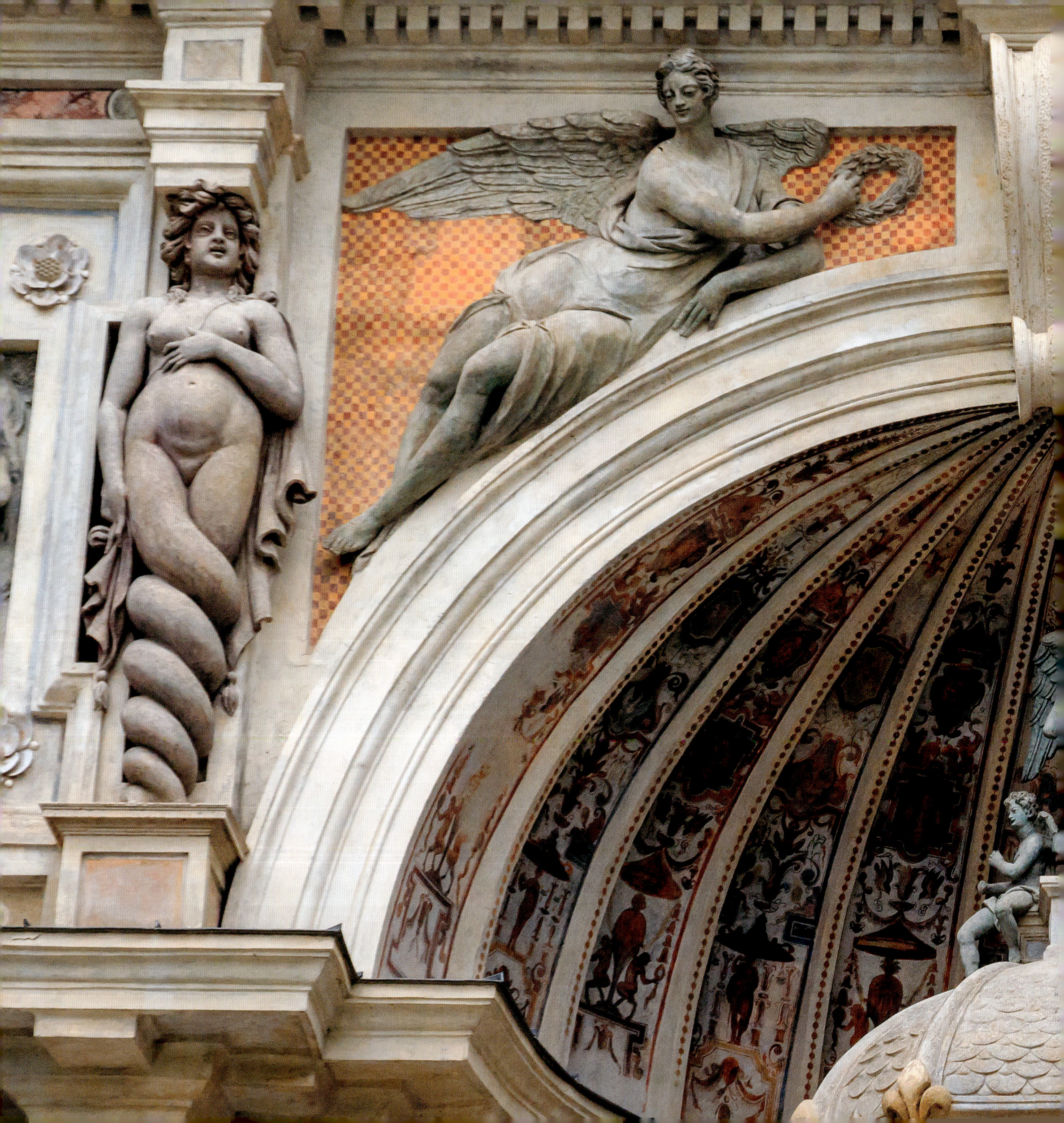

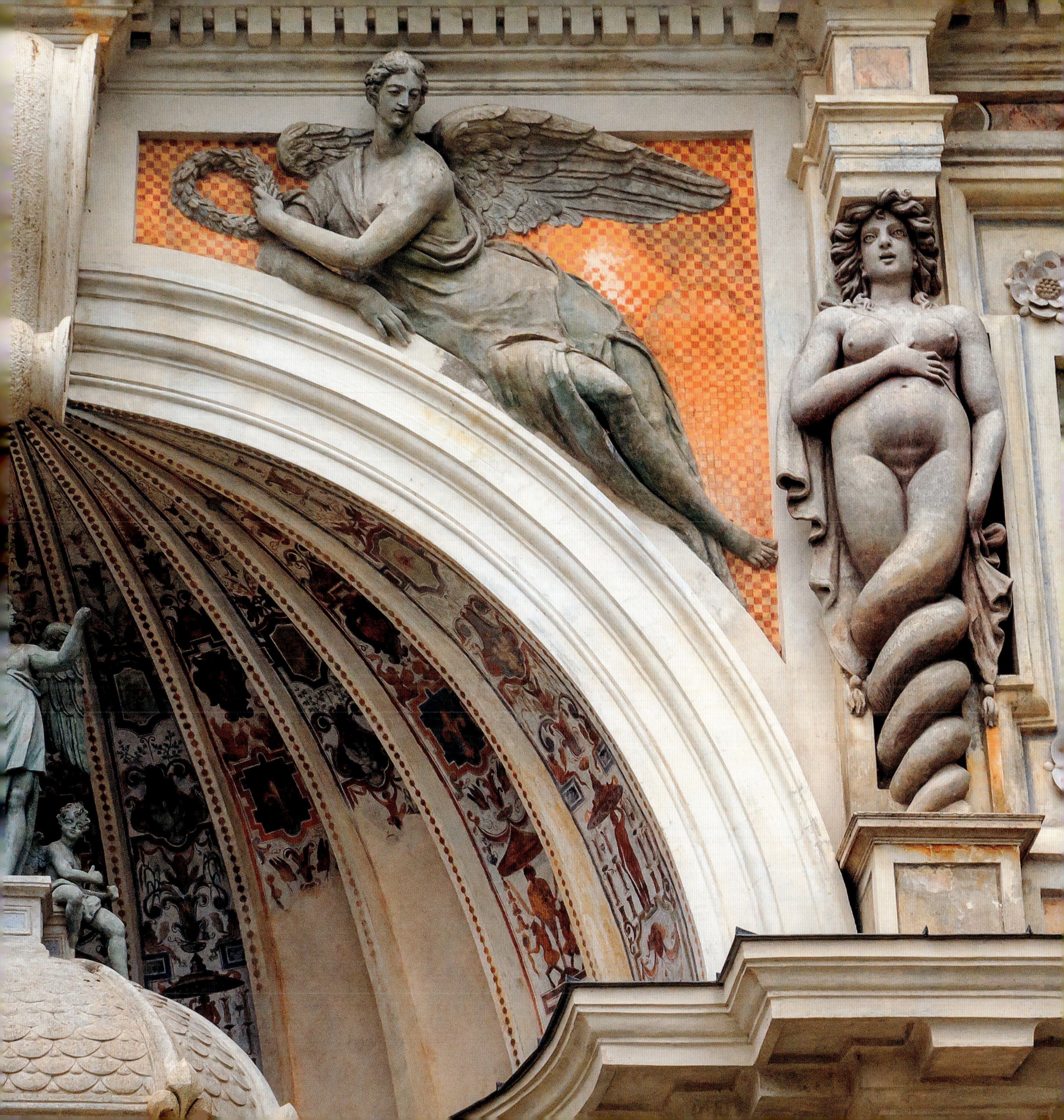

into a rectangular shaft in which a mixture of air and water was generated and introduced into a vertical pipe to reach a watertight chamber (aeolian chamber or wind chamber). The forceful injection of water caused the air to rise while the water was conveyed towards a water wheel to rotate a toothed roller that allowed the valves of the pipes of an organ housed inside the apse of the fountain with the instrument immersed in the water to open and close. The system was so deteriorated by the seventeenth century that Cardinal Alessandro ordered replacing it with a similar one. The Aeolian chamber and basin, the only elements remaining from the original apparatus, have been repurposed and restored using non-perishable materials.

Guests at Villa d'Este were so astounded by this device that it was quickly imitated and included in other lavish villas. Of these, only the one at Villa d'Este and the one at the Quirinale remain functional today. From the terrace, besides the external view, you can admire the Fish Pools, and on the enclosure wall, you'll find the Fountain of Diana Ephesia or Mother Nature. This was built in 1611 after installing a statue formerly at the Fountain of the Organ in its current position. At the time, the fountain was called the Fountain of the Flood (because of the waterfall below) or the Fountain of Mother Nature because of the 1568 sculpture by Gillis van den Vliete, inspired by the famous iconography of the Ephesian Artemis, an example of which had been found in Hadrian's Villa. The monumental elevation of the Fountain of the Organ forms the *scaenae frons* of the water spectacle designed by Ligorio. Following the seventeenth-century interventions, the decorative apparatus, hitherto highlighted by only four telamons (a male figure in the form of a pillar), was enriched with stucco statues of Apollo and Orpheus in the niches on the first order, mermaid-shaped Caryatids in the second, winged Victories in the spandrels at the keystone of the archway, and stucco bas-reliefs with mythological scenes relating to music. The basin was built, while inside the niche, the statue of Mother Nature was moved, and the octagonal aedicula was created to house the organ. A snippet of information from the first

Fountain of Ephesian Artemis or Mother Nature, early seventeenth century, reusing the statue of Mother Nature sculpted by Gillis van den Vliete in 1568

The statue was originally commissioned to the Flemish sculptor to adorn the central niche of the Fountain of the Organ. Alessandro d'Este had it moved in 1611 to its current more marginal position, perhaps so as not to contravene the dictates imposed by the Counter-Reformation which condemned works with pagan subjects. On a background decorated with tiburtine tartar, a material of limestone concretion widely used in the decorations of the lower part of the villa, is a travertine copy of the Diana of Ephesus, with her many breasts (according to some scholars, these are in fact scrotums of a bull, an animal linked to the cult of the goddess), symbolising the fertility of nature and the uninterrupted flow of life.

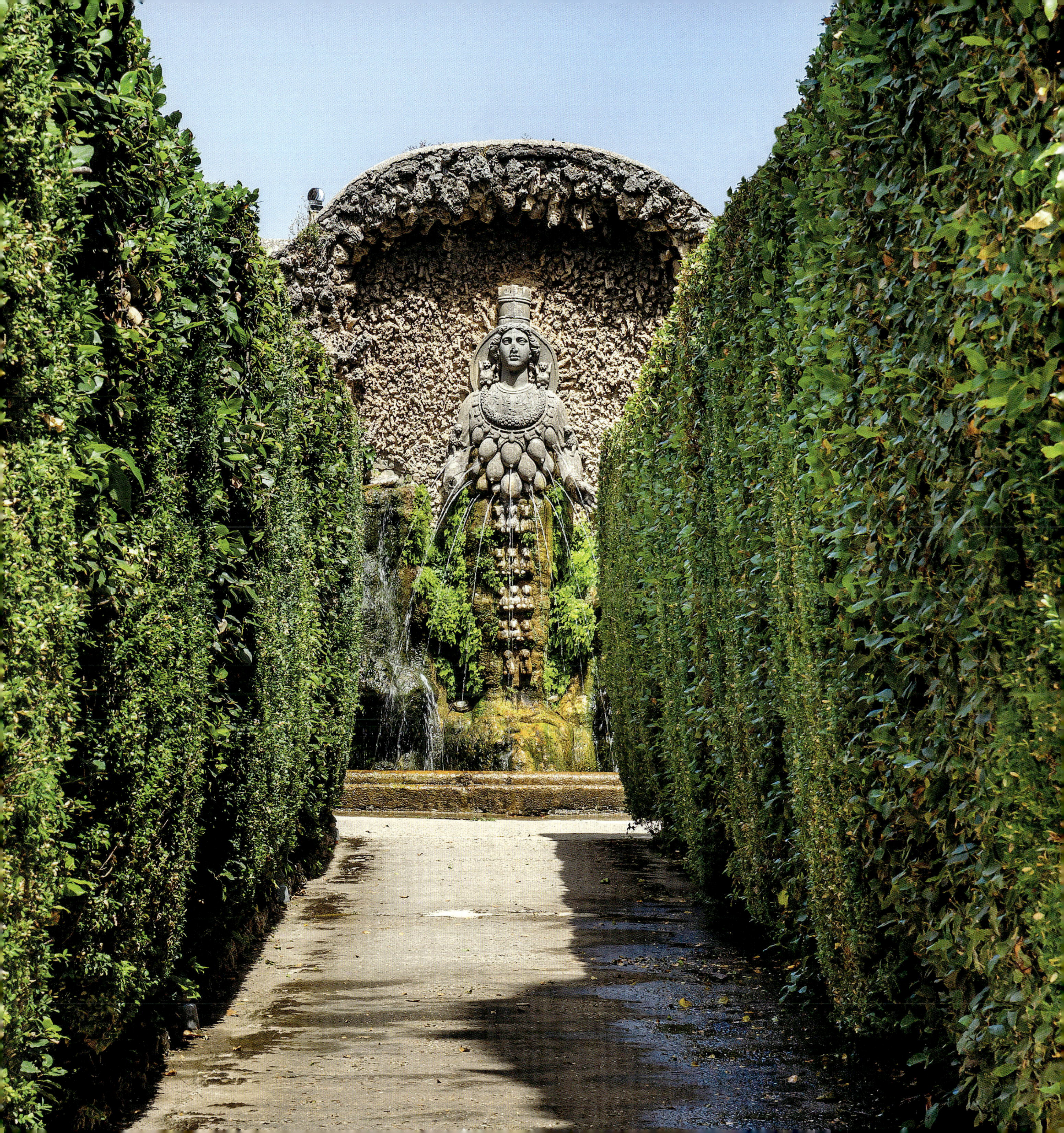

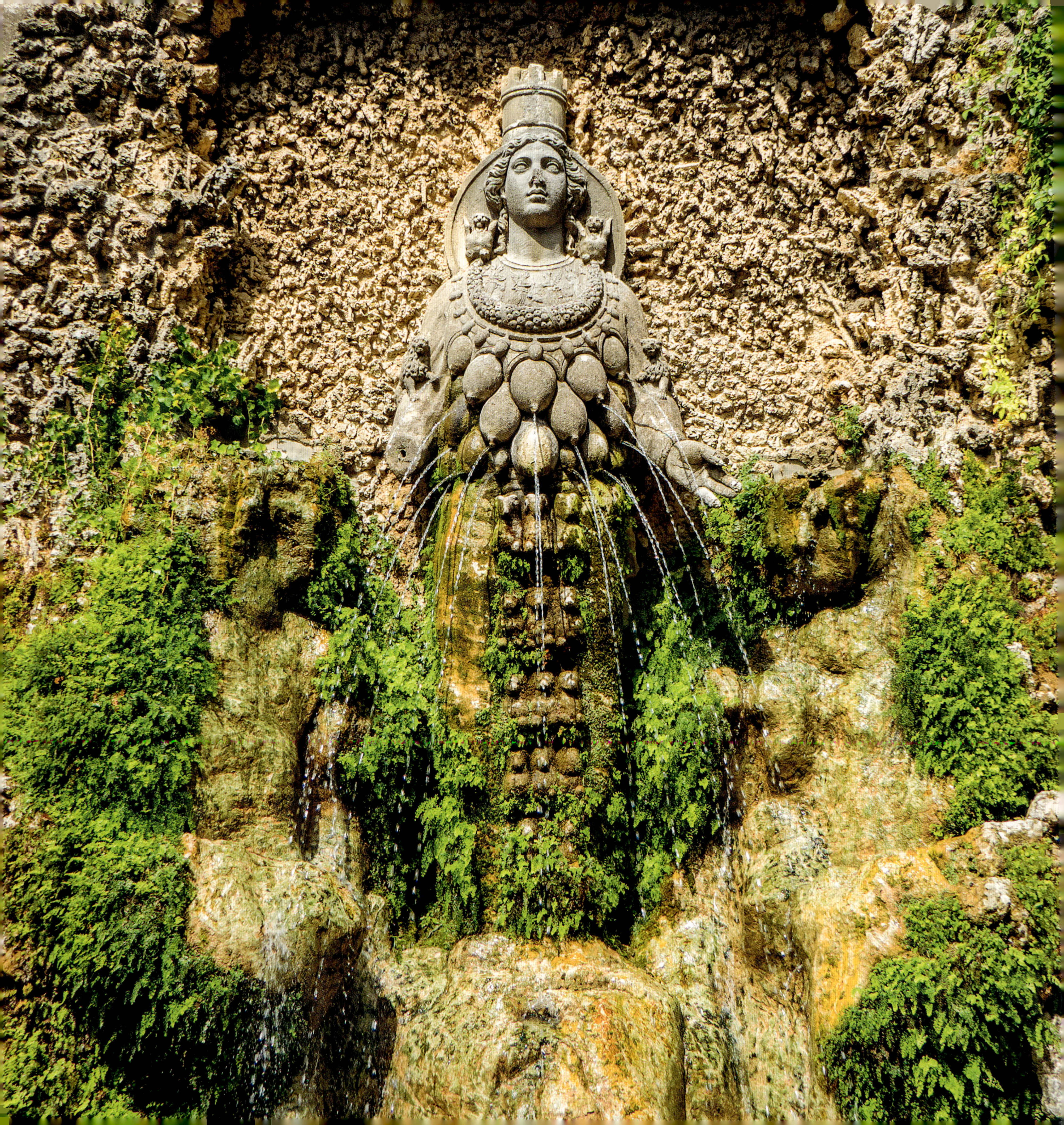

decades of the seventeenth century indicates that an unidentified portion of the wall next to the fountain was home to one of the many paintings that decorated the interior of the garden enclosure. While one can imagine from this information the extent of the luxury and splendour throughout the garden, including the perimeter walls, the subject depicted has itself particular significance, given that it could derive from a Ligorio original and given, moreover, that it must have constituted a dialogue and counterpoint with the Rometta Fountain. The document describes it as "Rome depicted all in line and painting, likenesses and letters and of other temple adornments".